ELECTRICAL PIONEERS OF AMERICA, THEIR OWN WORDS

Bell, De Forest, Edison, Franklin, Henry, Steinmetz, Tesla, Thomson, and Westinghouse

Edited by Stephen P. Tubbs, M.Sc., P.E.

Volume arrangement, introduction, and biographies copyright 1997
By Stephen Tubbs
1344 Firwood Dr.
Pittsburgh, PA 15243

Type reset and new cover in 2003.

No part of the material of this book may be reproduced commercially by offset-lithographic or equivalent copying devices without the permission of the publisher.

ISBN 0-9659446-2-X

CONTENTS

                                          PAGE

Introduction... v

Alexander Graham Bell
DISCOVERY AND INVENTION... 1
TO THE CAPITALISTS OF THE ELECTRIC TELEPHONE COMPANY............. 12
IMPROVEMENT IN TELEGRAPHY <Patent>... 15

Lee de Forest
HOW I INVENTED THE AUDION... 25

Thomas Alva Edison
MAN AND MACHINE.. 35
THE INVENTOR'S LOT... 38
THEY WON'T THINK.. 39
THEY DO WHAT THEY LIKE TO DO... 42
MACHINE AND PROGRESS.. 43
THE DESIRE FOR CHANGE.. 47
AGE AND ACHIEVEMENT.. 48
ELECTRIC LAMP <Patent>... 49
IMPROVEMENT IN PHONOGRAPH OR SPEAKING MACHINES <Patent>...... 54

Benjamin Franklin
THE KITE EXPERIMENT.. 59
OF LIGHTNING, AND THE METHOD (NOW USED IN AMERICA) OF
SECURING BUILDINGS AND PERSONS FROM ITS MISCHIEVOUS EFFECTS..... 61

Joseph Henry
ON THE PRODUCTION OF CURRENTS AND
SPARKS OF ELECTRICITY FROM MAGNETISM.................................... 64

Charles Proteus Steinmetz
THE EDUCATION OF ELECTRICAL ENGINEERS................................... 71
ELECTRICITY AND CIVILIZATION.. 76

Nikola Tesla
TRANSMISSION OF ELECTRICAL ENERGY WITHOUT WIRES................. 85
MY INVENTIONS; THE DISCOVERY
OF THE TESLA COIL AND TRANSFORMER.. 91
APPARATUS FOR TRANSMITTING ELECTRICAL ENERGY <Patent>......... 101

Elihu Thomson
WHAT IS ELECTRICITY?.. 107

George Westinghouse
ELECTRICITY IN THE DEVELOPMENT OF THE SOUTH.......................... 135

INTRODUCTION

This book is a compilation of writings of great American electrical pioneers at the dawn of the electrical age.

They were a diverse group. Many were not born in the U.S., but became naturalized U. S. citizens. They came from different economic levels. Edison started out poor, went to work young, and was practically self-educated. Westinghouse had the benefits of wealthy parents and a good education. They had varied interests. Electricity and inventing were not the main interests of Benjamin Franklin. Nikola Tesla, on the other hand, was almost totally devoted to electrical engineering, inventing, and experimenting. However, all were self-motivated, hard workers, inventive, and intelligent.

Minimal biographical information is supplied. The goal of this book is to present the words of these pioneers, not to present a biographer's view point.

All original text material has been retyped. On the rare occasions that spelling errors were encountered in the original material, the errors were corrected. Otherwise, the words presented are the same as originally published.

ALEXANDER GRAHAM BELL was born in Edinburgh, Scotland in 1847 and died in Beinn Bhreagh, Nova Scotia in 1922. He became a naturalized U.S. citizen in 1882. Most famous for his invention of the telephone in 1876, he also produced inventions in many other areas. He was the sole author of 18 patents and a collaborative author on 12 others. Some of these were in telegraphy, phonographs, metal detectors, flying machines, and hydrofoil boats. He formed the Bell Telephone Company and founded the <u>Science</u> magazine, the American Association to Promote Teaching of Speech to the Deaf and the Aerial Experiment Association.

Following is a speech given by Alexander Graham Bell to the graduating class of the Friend's School, Washington, D. C., delivered May 22, 1914. It was printed in the <u>National Geographic Magazine</u>, June 1914, Vol. XXV, p. 649-55.

"DISCOVERY AND INVENTION"

I am going to begin tonight by asking you a rather startling question: Did you ever put your head under water and chuck two stones together to see what the sound is like? If you have never done that, try it, and you'll get a new sensation. I did it once, and it sounded as if a man were hammering for all he was worth at my very ear.

I then took two tiny little pebbles and tapped them together quite lightly under water, and it sounded like a man knocking at the door. It was rather startling to hear such a loud noise from such a slight cause.

Of course, the question at once arose in the mind: How far off could we hear the sound? So I sent a boy a hundred feet up the beach with a couple of stones, directing him to strike them together under water. I then submerged my head, and I could hear the sound about as readily as before.

Well, I determined to test the maximum possible distance, and sent the boy across the bay in a boat to the other side, to a point at least a mile away from the place where I stood, and I followed him with field glasses to see that he carried out my instructions. I saw him land on the other side, take off his coat, roll up his sleeves, and go down to a little plank wharf on the shore rising only a few inches out of water. He lay down upon the wharf, face downward, and put his hands, into the water, and I then knew he was making signals with these stones.

Now, the question was: Could I hear him? Quietly and gently I went into the water at my side of the bay, submerged my head, and listened for all I was worth. Well, you know, the signals came perfectly clear and distinct, through more than a mile of water, to my ear. It was one of the most astonishing revelations of what could be done with water.

You know if you look away in the distance at a man firing a gun you can see the flash, and after a time you get the report; the sound takes time to travel through the air. It goes about 1,100 feet per second; but in the water it goes five times as fast as that--over 5,000 feet per second. Water is a much better conductor of sound than air.

DO FISH SIGNAL TO ONE ANOTHER BY SOUNDS?

Reflecting upon these various experiments, the thought occurred: If two little stones tapped together can be heard under water, why, every tiny lobster that snaps his claws must make an audible click. I wonder if there are creatures in the water that signal to one another by sound.

Well, I had occasion to try it once. Bathing in the Grand River in Ontario a great many years ago, I put my head very gently under water and listened, and, sure enough, "tick, tick," came a sound like a grasshopper's chirrup, and a little while after that a chirrup on the other side. There were creatures under the water that were calling to one another.

I don't know whether all fish make sounds or not, but there are some fish that certainly do. The drumfish on our coast drums away in the water so loudly that you can hear him while you are walking on the shore.

It is also a significant fact that all fish have ears. Why should they have ears if there is nothing for them to hear?

Of this we may be certain--that *there is a whole world of sound beneath the waves waiting to be explored*, perhaps by some of you.

I have wanted you to see how one observation leads to another. Starting with a very small thing--the chucking together of two pebbles under water, and following this up by other observations-we broaden our field of knowledge and reach generalizations of considerable magnitude as the resultant of numerous small thoughts brought together in the mind and carefully considered.

OUT OF THE BEATEN TRACK

I was walking along the road one day in my country place in Nova Scotia, when the idea occurred to leave the beaten track and dive into the woods. Well, I had not gone 50 feet before I came upon a gully, and down at the bottom was a beautiful little stream. I never knew of it before.

Of course, I was not satisfied with the mere discovery, but went down into the gully and explored it right and left. I followed it up to its source. I followed it downward for half a mile, through a beautiful moss-grown valley, until at last the little streamlet discharged into a pond, and away in the distance I could see a sea beach with the open water beyond.

Now, just think of that! Here was a beautiful gorge, half a mile long, right on my own place, and coming at one point within 50 feet of a well-trodden road, and I never knew of its existence before. We are all much inclined, I think, to walk through life with our eyes shut. There are things all round us and right at our very feet that we have never seen, because we have never really looked.

Don't keep forever on the public road, going only where others have gone and following one after the other like a flock of sheep. Leave the beaten track occasionally and dive into the woods. Every time you do so you will be certain to find something that you have never seen before. Of course it will be a little thing, but do not ignore it. Follow it up, explore all round it; one discovery will lead to another, and before you know it you will have something worth thinking about to occupy your mind. All really big discoveries are the results of thought.

THE BEGINNINGS OF INVENTION

I dare say you have all heard of that celebrated painter who would never allow any one to mix his colors for him. He always insisted on doing that himself, and at last one of his students, whose curiosity had been aroused, said: "Professor, what do you mix your colors with?" "With brains, sir," said the professor. Now, that is what we have to do with our observations.

I think I left you with your head under water listening to the clicking of two stones. Now, let us see whether we cannot use our brains to get you out of so awkward a predicament. We will then have entered the realm of invention, as distinct from discovery.

Why should we not simply put the ear to the water instead of submerging the whole head? Why should we not ring a bell under water instead of clicking stones together to make a noise. An ordinary dinner bell would do. Empty it of air and ring it under water, and the sound can be heard by a submerged ear at a great distance away.

It is a little awkward, however, to keep the ear continuously submerged on account of the movements of the surface water. Every now and then a little wave will slap you in the face, and you are apt to choke if you are caught unprepared.

Why would it not be better to transmit the sound vibrations from the water to the ear through some intervening mechanism, and thus obviate the necessity of submerging the ear at all?

I have tried submerged hearing tubes of various kinds and planks of wood partially submerged, with the ear applied to the part out of water.

If you put your ear to the bottom of a boat--inside, of course, not outside--you can hear a bell at a distance quite readily. It still is a little awkward, however, to get your ear against the planks of the boat; but brains will help you out. Just fix a telephone transmitter to the planks of the boat, and you can sit at ease with the telephone receiver at your ear.

You may even put the telephone transmitter overboard. It then becomes a submerged ear and will listen for you under water.

FISHING WITH TELEPHONES

I have often thought I should like to go on the banks of Newfoundland and fish with a telephone. If you were to send the transmitter down among the codfish with the bait, perhaps you would find something there to hear. I have never tried it. I will leave that to you.

We now have numbers of steamers upon the Atlantic fitted with telephone transmitters attached to the thin iron skin of the hull, away down in the hold, and the receiving telephone on the bridge.

On shore there are huge bells at lighthouse stations making fog-signals under water, and each steamer as it approaches the coast can pick up these submarine sounds at a distance of 10 miles.

Here is a completed invention which some patient observer has evolved from just such little beginnings as those I have described.

I doubt whether you could hear a fog-signal through the air at any such distance as that. The air is at best but a poor conductor of sound, and many illusions of hearing are possible.

It is difficult in any case to tell the exact direction of a sound in a fog. It is possible, too, that you might have an echo from the sails of a vessel, and you would then be entirely misled as to the direction of the signal station.

Then, again, an island anywhere near casts a sound-shadow upon the water. The sound-wave striking the island is deflected up into the sky, and you would have to go up in a balloon to hear it, and it may not come down again to the surface for a mile or two beyond the island. A ship quite close to the island might not hear the sound. The captain, knowing that the fog-horn should be heard at least a mile or two away, imagines himself to be much farther off than he really is, and in the midst of the fog he may become conscious of the presence of the land only by actual contact with it.

Then the transmitting qualities of the air are subject to variations on account of unusual atmospheric conditions. You may be quite near a fog-signal station and yet hear the sound so faintly that you imagine it to be far away. You may even get an echo from the clouds; but then you know you are subject to an illusion, for the sound would seem to come from the sky.

Now, sounds can be transmitted through the water to far greater distances than through the air, and atmospheric conditions have no effect.

I don't want to confine your attention to inventions that already have been made. I want to show you also that there is room for something new. We don't know everything yet and the list of possible inventions is not yet closed. Take, for example, the case we have been talking about, the transmission of sound through water.

EXPLORING UNDER THE SEA

Three-quarters of the earth's surface is under water and has not yet been explored, at least to any great degree. The only way we have of reaching the mountains and valleys at the bottom of the sea is by sending down a sounding line and bringing up a specimen of the bottom attached to the sinker. It is no joke, however, to reach the bottom of the deep, blue sea through one mile or even two miles of water, and it takes several hours to make a single sounding. Just think of all the time and labor involved in merely ascertaining the depth.

Why should we not send down a sound instead and listen for an echo from the bottom. Knowing the velocity of sound in water and the time taken for the echo to reach the ear, we should be able to ascertain the depth of the deepest part of the ocean in less than four seconds instead of more than four hours. Here is something worth doing. It has never been tried. I have suggested it a number of times, and I will now pass on the thought to you in the hope that some of you may care to take it up.

Suppose you are on one of those steamers provided with transmitter hulls and telephone ear-pieces, and you send down a little piece of gun-cotton or other explosive material to a safe distance below your ship and then explode it by an electrical contact. The sound-wave from the explosion will, of course, go down to the bottom and then be reflected up again, so that after a certain length of time you should get an echo from the bottom.

Not only should you be able to tell the depth of the ocean by an echo from the bottom, but you might perhaps learn something of the nature of the bottom itself. A flat bottom should yield a single sharp return, whereas an undulating bottom should yield a multiple echo, like that heard when you fire a pistol among hills.

Then, as you approach the shore you should get resonance effects, like those perceived when you shout out loudly in an empty cave.

However, I must not take up your time in speaking upon only one subject. What I want to direct your attention to is that both discovery and invention are not things that come all at once. They arise from very simple beginnings. A small observation, patiently followed up by other observations equally small, leads gradually to a big conclusion. Do not ignore little things; life itself is made up of them, and there is a good old Scotch saying that bears upon the point:

"Mony a mickle maks a muckle."

A great many small things make a big one. Any one, if he will only observe, can find some little thing he does not understand as a starter for an investigation.

AN EXPERIMENT AT HOME

I had rather a curious illustration of this the other day in my own house. I told a lad who was waiting upon me that I wanted to make some experiments with a bottle of water, and told him to bring a bottle of very hot water from the kitchen, and be sure that it was quite full. He soon returned with a big-bodied bottle provided with a long and narrow neck, filled to the brim, and put it on the mantelpiece and went downstairs. After the water had cooled, I rang the bell for John.

"John," I said, "I thought I told you to fill that bottle quite full."

"So I did, sir," he replied.

"Well, look at it now; it's not nearly full; the neck is quite empty."

John assured me that he had not touched the bottle since he first put it up, and I assured him that I had not poured any of the water out.

"Well," I said, "what has become of the water?"

He was quite nonplussed at first, and then he began to--to-ratiocinate, and said: "The water was quite hot when I put it in; there was steam coming from it. The water must have evaporated."

I made no comment, but looked at him and said: "Let's try it again. You fill that bottle chock full of hot water this time, and then *cork it* so that no steam can escape."

He did so; and by and by I rang the bell again, and up came John.

"John," I said, "I thought you filled that bottle quite full."

"So I did sir," he replied.

"Well look at it now; it's not nearly full." John assured me that he had not touched the cork, and I replied: "Well, what has become of the water?" John said he didn't know. He admitted that some of it had evidently gone, but where it had gone he couldn't for the life of him conceive, and he hasn't found out yet.

I am sorry now that I didn't think of telling John to weigh the bottle when he first brought it up, for by weighing it again he could have found out exactly how much had disappeared.

If John hadn't given up he might have arrived by degrees at a realization of the principle upon which a thermometer works.

A thermometer is an instrument for measuring heat, and whenever you can measure a phenomenon you have a basis upon which may be built a science; in fact, all science is dependent upon measurement.

When you measure heat you get the science of thermo-dynamics, and thermo-this and thermo-that. When you measure the pressure of the atmosphere by a barometer you lay the basis for the science of meteorology and a whole lot of sciences dependent upon atmospheric measurements. So you have sciences based upon the measurement of sound and light; but you have no science of odor.

MEASURING AN ODOR

Did you ever try to measure a smell? Can you tell whether one smell is just twice as strong as another? Can you measure the difference between one kind of smell and another. It is very obvious that we have very many different kinds of smells, all the way from the odor of violets and roses up to asafetida. But until you can measure their likenesses and differences you can have no science of odor. If you are ambitious to found a new science, measure a smell.

What is an odor? Is it an emanation of material particles into the air, or is it a form of vibration like sound? If you can decide that, it might be the starting point for a new investigation. If it is an emanation, you might be able to weigh it; and if it is a vibration, you should be able to reflect it from a mirror. You can reflect sound and light and heat, and I have even warmed my hands at the reflection of a fire in a mirror. Not a glass mirror, for glass is opaque to radiant heat. A sheet of transparent glass makes a fine fire-screen. You can see the fire through it, but it cuts off the heat. When you try to reflect it from an ordinary looking-glass, the heat has to go through the glass in order to reach the reflecting surface behind and then pass through the glass a second time in order to get out. Take a sheet of polished metal--tin-foil will do--or any metal with a bright and shiny surface and you can reflect heat from it with ease.

Can you reflect a smell or measure its velocity of transmission? If you can do those things you will be well advanced on the road to the discovery of a new science.

THE SMELL OF TELLURIUM

Well, that reminds me of a discovery that started with a smell. We have a very rare elementary substance known as tellurium, and when you melt it with a blow-pipe it gives off a smell. We can't measure it, nor even describe it; but if you have ever smelled it you will know it ever after. There is nothing in heaven or on earth that smells like that.

Now, you know it is the object of many chemists and scientific men to turn their discoveries to some practical use. Then try, through chemical and other means, to convert waste products, for example, into useful things. Indeed, the utilization of waste products is a characteristic of the age in which we live.

Just think of what they have done. Here is a gas manufactory consuming coal. After the gas has been produced we have left upon our hands ashes and clinkers and a lot of evil-smelling tar. Well, the chemists go to work and out of that tar they make the most delightful perfumes for scenting handkerchiefs, and nice sweet essences for flavoring puddings, and the most beautifully colored dyes, all made from coal-tar.

Now, there was a distinguished chemist who thought he saw a chance of making something valuable out of waste products obtained in the manufacture of sulphuric acid. Some of the powder he obtained he heated with a blow-pipe, and at once perceived the characteristic smell of tellurium. Here, he thought, was a rare and valuable element contained in a common and cheap by-product and it might pay to extract it. He then applied various chemical tests, but could get no other indication of the presence of tellurium excepting the smell. All the reactions declared there was no tellurium there.

He did not stop with this observation, but followed it up and began reasoning about it. If, he thought, there is no tellurium here, there is certainly something that has a smell very like it, and I know of no other substance on earth that has a smell like that. Perhaps there may be a new substance here, not yet discovered, which resembles tellurium, at least in the smell.

He knew that he was working with a regular conglomerate or mixture of all sorts of materials, many of which he could identify. He then extracted from the mass all the materials he knew were there to see if there was anything left; and, sure enough, a residue appeared which turned out to be, as he had suspected, a new elementary substance not heretofore known to man.

SELENIUM FOUND

He termed this substance *selenium* because it resembled tellurium. The word selenium, you know, is derived from a Greek word meaning the moon, and tellurium comes from the Latin--tellus, the earth. The two substances were not identical, but were related to one another as the moon is to the earth.

Selenium was found to resemble black sealing-wax in appearance. It had a beautiful, black, glossy surface, and in thin films was transparent, showing ruby red by transmitted light. In this, its vitreous form, it was a non-conductor of electricity, thus differing in a remarkable degree from tellurium, which was a good conductor.

When, however, selenium was heated almost to the fusing point and then allowed to cool very slowly, it completely changed its appearance. It acquired a dull metallic look, like lead; and in this, its crystalline condition, was also found to be a conductor of electricity, but of extremely high resistance. A little pencil of crystalline selenium not much more than an inch in length offered as much resistance to the passage of an electrical current as 96 millions of miles of wire, enough to reach from here to the sun, and yet it was a conductor. That was a discovery. Now comes an invention.

Willoughby Smith, in laying the Atlantic cable, found it advisable to balance the electrical resistance of the cable during the process of submersion by tremendous coils of well-insulated wire. Why, thought he, should not a little bit of selenium balance the whole cable and enable us to get rid of all this complication of wire.

He succeeded in doing this, but found the electrical resistance very variable. At times the selenium would balance the whole cable and at other times not one-half of it.

He did not stop with this observation, but sought the cause of the variation. He multiplied observations, and his assistant, Mr. May, soon discovered that the resistance of the selenium was greater at night than in the day.

This at once suggested to Willoughby Smith the thought that perhaps the electrical resistance of selenium was affected by light, and he proceeded to put his idea to the test of experiment. He shut up the selenium in a dark box near a bright light, and found that when the lid was open the resistance went down and when it was closed it rose again. Even a shadow falling upon the selenium affected its electrical resistance.

SPEECH FROM A SUNBEAM

Then other scientific men took the matter up. Professor Adams, of King's College, England, discovered that the resistance varied directly with the intensity of the light that fell upon the selenium. Then I came along with some speculations concerning the possibilities of telephoning without wires by varying the intensity of a beam of light by the action of the voice, and allowing the light to fall upon a piece of crystalline selenium. In this way I thought it would be possible to get speech from a sunbeam.

Well, I need not go into the details, but it was true. I produced the *photophone*, an instrument for talking along a beam of light instead of a telegraph wire. It is interesting to remember that all these things resulted from the observation of a smell.

When I was invited to talk to you tonight I had no idea of what to say. I thought of all the good maxims for your future conduct in life; but giving advice to young people is out of my line, and it seemed to be better to choose some subject with which I was a little familiar myself.

How discoveries and inventions arise from the observation of little things is surely a topic worthy of your consideration. I also thought it would be interesting for you to know how many apparently impossible results have been actually achieved by the patient multiplication of little observations.

It was only a short time ago that if you wished to express the idea that anything was utterly impossible you would say, "I could no more do that than I could fly." I don't think there is any one here who is too young to have heard that expression. It was the height of impossibility that we should fly, and here men are flying in the air today.

It is only a few years since the first man flew, and we are only at the beginning of aviation. What a delightful idea it is to go sailing through the air. The only trouble is that you must come down, and we have altogether too many fatalities connected with the work. Here, then, is a subject for you to explore: How to improve the safety of the flying machine. How to produce flying machines that any one can fly.

We know perfectly well that the time is coming, and is almost here, when it will be an every-day thing to go from place to place through the air. Perhaps some of you may find a field of occupation in bringing this about.

FLYING ACROSS THE ATLANTIC

Even today we have startling propositions to do things that are apparently impossible. A man proposes to try this summer to fly across the Atlantic Ocean in a heavier-than-air flying machine. The strange thing about the matter is that experts who have examined into the possibilities find that he really has a fighting chance.

You see the distance is less than 2,000 miles from Newfoundland to Ireland. This means that if you could go at 100 miles an hour you would cross the Atlantic in 20 hours--less than a day. Just think of that. Well, we have flying machines that go at a greater speed than that. We already have machines that could cross the ocean if their engines can keep going for 20 hours.

Of course, these are exceptional machines; but even the ordinary machines of today make 50 miles an hour with ease. Now, a flying machine flies faster as you go higher up, because the rarer air offers less resistance to the motion, while the propeller gives the same push with the same power, whatever the elevation. As you get into rarer air the propeller simply spins round faster.

A 50-mile-an-hour machine flying two miles high in the air-and we have machines that have gone twice as high as that--will fly much faster than 50 miles an hour.

Then at an elevation of two miles high in the air there is a constant wind blowing in the general direction of Europe having a velocity anywhere from 25 to 50 miles an hour.

As the net result of all these things, there can be little doubt that any ordinary machine that is able to support itself in the air at an elevation of two miles high will attain a speed of at least 100 miles an hour in the direction of Europe, and that means going from America to Europe in a single day.

Calculation shows that, taking all these circumstances into consideration, our best machines should be able to cross the Atlantic in 13 hours. I hardly dare to say it aloud for publication. It is sufficiently startling to know that it is not only possible, but probable, that the passage may be made in a single day. But if, as I imagine, it can be done in 13 hours, you may take an early breakfast in Newfoundland and a late dinner in Ireland the same night.

Now, I will not take up any more of your time. My idea has been to point out to you how many great discoveries and inventions have originated from very little things, and to impress upon your minds the importance of observing closely every little thing you come across and of reasoning upon it.

Indeed as Smiles very happily puts it, "The close observation of little things is *the secret of success* in business, in art, in science, and in every pursuit in life."

Following is a letter written by Alexander Graham Bell to "The Capitalists of the Electric Telephone Company" from Kensington on March 25, 1878. It can be found reprinted in the book by Waite, H. E., <u>Make a Joyful Sound</u>, (Philadelphia: Macrae Smith Company, 1961), p.265-70.

"TO THE CAPITALISTS OF THE ELECTRIC TELEPHONE COMPANY"

Gentlemen,

It has been suggested that at this our first meeting I should lay before you a few ideas concerning the future of the Electric Telephone together with any suggestions that occur to me in regards to the best mode of introducing the instrument to the public.

The Telephone may be briefly described as an electrical contrivance for reproducing in distinct places the tones and articulations of a speaker's voice so that conversation can be carried on by word of mouth between persons in different rooms, in different streets or in different towns.

The great advantage it possesses over every other form of Electrical apparatus consists with the fact that it requires no skill to operate the instrument. All other telegraphic machines produce signals which require to be translated by experts and such instruments are therefore extremely limited with their applications but the Telephone actually <u>speaks</u> and for this reason it can be utilized for nearly every purpose for which speech is employed.

The chief obstacle to the universal use of electricity as a means of communication between distant points has been the skill required to operate Telegraphic instruments. The invention of Automatic Printing Telegraphs, Dial Instruments, etc. has materially reduced the amount of skill required but has introduced a new element of difficulty in the shape of increased expense. Simplicity of operation has been obtained by complication of the parts of the machine--so that such instruments are much more expensive than those usually employed, by skilled electricians. The simple and inexpensive nature of the Telephone on the other hand renders it possible to connect every man's house or manufactory with a central station so as to give him the benefit of direct Telephonic Communication with his neighbors at a cost not greater than that incurred for gas or water.

At the present time we have a perfect network of gas pipes and water pipes throughout our larger cities. We have main pipes laid under the streets communicating by side pipes with the various dwellings enabling the inmates to draw their supply of gas and water from a common source.

In a similar manner, it is conceivable that cables of Telephonic wires could be laid underground or suspended overhead communicating by branch wires with private dwellings, counting houses, shops, manufactories, etc. uniting them through the main cable with a central office where the wires could be connected together as desired establishing direct communication between any two places in the City. Such a plan as this though impracticable at the present will, I firmly believe, be the outcome of the introduction of the Telephone to the public--not only so but I believe that in the future wires will unite the head offices of Telephone Companies with different cities and a man in one part of the Country may communicate by word of mouth with another in a distant place. I am aware that such ideas may appear to you Eutopian and out of place for we are met together for the purpose of discussing not the future of the Telephone but its present. Believing however as I do that such a scheme will be the ultimate result of the introduction of the Telephone to the public I could impress upon you all the advisability of keeping this end in view that all present arrangements of the Telephone may eventually be utilized in this grand system.

I would therefore suggest that in introducing the Telephone to the Public at the present time you should on no account allow the Control of the Conducting Wires to pass out of your hands. The plan usually pursued in regard to private Telegraphs is to lease such Telegraph lines to private individuals or to Companies at a fixed annual rental. This plan should be adopted by you but instead of erecting a Line directly from one point to another I would advise you to bring the Wires from the two points to the Office of the Company and there connect them together. If this plan be followed a large number of Wires would soon be controlled by the Telephone Office where they would be easily accessible for testing purposes. In places remote from the Office of the Company simple testing boxes could be erected for the Telephone wires of that Neighborhood and these testing places could at anytime be connected into central offices when the Lessee of the Telephone Wires desire inter-communication.

In regard to other present uses for the Telephone the Instrument can be supplied so cheaply as to compete upon favorable terms with speaking tubes, bells and annunciators as a means of communication between different parts of a House.

This seems to be a very valuable application of the Telephone not only on account of the large number of Telephones that would be wanted but because it would lead eventually to the plan of intercommunication referred to above--I would therefore recommend that special arrangements should be made for the introduction of the Telephone into Hotels and private buildings in place of the speaking tubes & annunciators at present employed.

Telephones sold for this purpose could be stamped or numbered in such a way as to distinguish them from those employed for business purposes and an Agreement could be signed by the purchaser that the Telephone should become forfeited to the Company if used for other purposes than there specified in the agreement.

It is probable that such a use of the Telephone would speedily become popular and that as the public became accustomed to the telephone in their house they would recognize the advantages of a system of inter communication. When this time arrives I would advise the Company to place Telephones free of charge for a specified period in a few of the principle shops so as to offer to those householders who work with the Central Office the additional advantages of oral communication with their trades people. The Central Office system once inaugurated in this manner would inevitably grow to enormous proportions for those shop keepers would naturally obtain the custom of such householders. Other shop keepers would thus be induced to employ the Telephone and as such Connections with the Central Office increased in number so would the advantage to householders become more apparent and the number of subscribers be increased. Should this plan be adopted the Company should employ a man in each Central Office for the purpose of connecting the Wires as directed. A fixed annual rental could be charged for the use of the Wires or a toll could be levied. As all connections would necessarily be made at the Central Office it would be easy to note the time during which any Wires were connected and to make a charge accordingly. Bills could be sent in periodically. However small the rate of charge might be the revenue would probably be something enormous.

In conclusion I would say that it seems to be that the telephone should immediately be brought prominently before the public as a means of communication between Bankers, Merchants, Manufacturers, Wholesale & retail dealers, Dock Companies, Gas Companies, Water Companies, Public Offices, Fire Stations, Newspaper Offices, Hospitals and Public Buildings and for use in Railway Offices, in Mines and in Diving Operations.

Arrangements should also be speedily concluded for the use of the Telephone in the Army and Navy and by the postal Telegraph Department.

Although there is a great field for the Telephone in the immediate present I believe there is a still greater in the future.

By bearing in mind the great objects to be ultimately achieved I believe that the telephone Company can not only secure for itself a business of the most remunerative kind but also benefit the public in a way that has never previously been attempted.

                I am Gentlemen
                    Your obedient Servant
                        Alexander Graham Bell

Following is a copy of Alexander Graham Bell's first telephone patent.

UNITED STATES PATENT OFFICE

ALEXANDER GRAHAM BELL, OF SALEM, MASSACHUSETTS

IMPROVEMENT IN TELEGRAPHY

SPECIFICATION forming part of Letters Patent No. 174,465, dated March 7, 1876

Application filed February 14, 1876

To all whom it may concern:

Be it known that I, ALEXANDER GRAHAM BELL, of Salem, Massachusetts, have invented certain new and useful Improvements in Telegraphy, of which the following is a specification:

In Letters Patent granted to me April 6, 1875, No. 161,739, I have described a method of, and apparatus for, transmitting two or more telegraphic signals simultaneously along a single wire by the employment of transmitting-instruments, each of which occasions a succession of electrical impulses differing in rate from the others; and of receiving-instruments, each tuned to a pitch at which it will be put in vibration to produce its fundamental note by one only of the transmitting-instruments; and of vibratory circuit-breakers operating to convert the vibratory movement of the receiving-instrument into a permanent make or break (as the case may be) of a local circuit, in which is placed, a Morse sounder, register, other telegraphic apparatus. I have also therein described a form of autograph-telegraph based upon the action of the above-mentioned instruments.

In illustration of my method of multiple telegraphy I have shown in the patent aforesaid, as one form of transmitting-instrument, an electro-magnet having a steel-spring armature, which is kept in vibration by the action of a local battery. This armature in vibrating makes and breaks the main circuit, producing an intermittent current upon the line-wire. I have found, however, that upon this plan the limit to the number of signals that can be sent simultaneously over the same wire is very speedily reached; for, when a number of transmitting-instruments, having different rates of vibration, are simultaneously making and breaking the same circuit, the effect upon the main line is practically equivalent to one continuous current.

In a pending application for Letters Patent, filed in the United States Patent Office February 25, 1875, I have described two ways of producing the intermittent current--the one by actual make and break of contact, the other by alternately increasing and diminishing the intensity of the current without actually breaking the circuit. The current produced by the latter method I shall term, for distinction sake, a pulsatory circuit.

My present invention consists in the employment of a vibratory or undulatory current of electricity in contradistinction to a merely intermittent or pulsatory current, and a method of, and apparatus for, producing electrical undulations upon the line wire.

The distinction between an undulatory and a pulsatory current will be understood by considering that electrical pulsations are caused by sudden or instantaneous changes of intensity, and that electrical undulations result from gradual changes of intensity exactly analogous to the changes in the density of air occasioned by simple pendulous vibrations. The electrical movement, like the serial motion, can be represented by a sinusoidal curve or by the resultant of several sinusoidal curves.

Intermittent or pulsatory and undulatory currents may be of two kinds, accordingly as the successive impulses have all the same polarity or are alternately positive and negative.

The advantages I claim to derive from the use of an undulatory current in place of a merely intermittent one are, first, that a very much larger number of signals can be transmitted simultaneously on the same circuit; second, that a closed circuit and single main battery may be used; third, that communication in both directions is established without the necessity of special induction-coils; fourth, that cable dispatches may be transmitted more rapidly than by means of an intermittent current or by the methods at present in use; for, as it is unnecessary to discharge the cable before a new signal can be made, the lagging of cable signals is prevented; fifth, and that as the circuit is never broken a spark-arrester becomes unnecessary.

It has long been know that when a permanent magnet is caused to approach the pole of an electro-magnet a current of electricity is induced in the coils of the latter, and that when it is made to recede a current of opposite polarity to the first appears upon the wire. When therefore, a permanent magnet is caused to vibrate in front of the pole of an electro-magnet an undulatory current of electricity is induced in the coils of the electro-magnet, the undulations of which correspond, in rapidity of succession, to the vibrations of the magnet, in polarity to the direction of its motion, and in intensity to the amplitude of its vibration.

That the difference between an undulatory and an intermittent current may be more clearly understood I shall describe the condition of the electrical current when the attempt is made to transmit two musical notes simultaneously--first upon the one plan and then upon the other. Let the interval between the two sounds be a major third; then their rates of vibration are in the ratio of 4 to 5. Now, when the intermittent current is used the circuit is made and broken four times by one transmitting-instrument in the same time that five makes and breaks are caused by the other. A and B, Figs. 1, 2, and 3, represent the intermittent currents produced, four impulses of B being made in the same time as five impulses of A. $c\ c\ c$, &c., show where and for how long time the circuit is made, and $d\ d\ d$, &c., indicate the duration of the breaks of the circuit. The line A and B shows the total effect upon the current when the transmitting-instruments for A and B are caused simultaneously to make and break the same circuit. The resultant effect depends very much upon the duration of the make relatively to the break. In Fig. 1 the ratio is as 1 to 4; in Fig. 2, as 1 to 2; and in Fig. 3 the makes and breaks are of equal duration. The combined effect, A and B, Fig. 3, is very nearly equivalent to a continuous current.

When many transmitting-instruments of different rates of vibration are simultaneously making and breaking the same circuit the current upon the main line becomes for all practical purposes continuous.

Next, consider the effect when an undulatory current is employed. Electrical undulations, induced by the vibration of a body capable of inductive action, can be represented graphically, without error, by the same sinusoidal curve which expresses the vibration of the inducing body itself, and the effect of its vibration upon the air; for, as above stated, the rate of oscillation in the electrical current corresponds to the rate of vibration of the inducing body--that is, to the pitch of the sound produced. The intensity of the current varies with the amplitude of the vibration--that is, with the loudness of the sound; and the polarity of the current corresponds to the direction of the vibrating body--that is, to the condensations and rarefactions of air produced by the vibration. Hence, the sinusoidal curve A or B, Fig. 4, represents, graphically, the electrical undulations induced in a circuit by the vibration of a body capable of inductive action.

The horizontal line $a\ d\ e\ f$, &c., represents the zero current. The elevations $b\ b\ b$, &c., indicate impulses of positive electricity. The depressions $c\ c\ c$, &c., show impulses of negative electricity. The vertical distance $b\ d$ or $c\ f$ of any portion of the curve from the zero-line expresses the intensity of the positive or negative impulses at the part observed, and the horizontal distance $a\ a$ indicates the duration of the electrical oscillation. The vibrations represented by the sinusoidal curves B and A, Fig. 4, are in the ratio aforesaid, of 4 to 5--that is, four oscillations of B are made in the same time as five oscillations of A.

The combined effect of A and B, when induced simultaneously on the same circuit is expressed by the curve A + B, Fig. 4, which is the algebraical sum of the sinusoidal curves A and B. This curve A + B also indicates the actual motion of the air when the two musical notes considered are sounded simultaneously. Thus, when electrical undulations of different rates are simultaneously induced in the same circuit, an effect is produced exactly analogous to that occasioned in the air by the vibration of the inducing bodies. Hence, the coexistence upon a telegraphic circuit of electrical vibrations of different pitch is manifested, not by the obliteration of the vibratory character of the current, but by peculiarities in the shapes of the electrical undulations, or, in other words, by peculiarities in the shapes of the curves which represent those undulations.

There are many ways of producing undulatory currents of electricity, dependent for effect upon the vibrations or motions of bodies capable of inductive action. A few of the methods that may be employed I shall here specify. When a wire, through which a continuous current of electricity is passing, is caused to vibrate in the neighborhood of another wire, an undulatory current of electricity is induced in the latter. When a cylinder, upon which are arranged bar-magnets, is made to rotate in front of the pole of an electro-magnet, an undulatory current of electricity is induced in the coils of the electro-magnet.

Undulations are caused in a continuous voltaic current by the vibration of the motion of bodies capable of inductive action; or by the vibration of the conducting-wire itself in the neighborhood of such bodies. Electrical undulations may also be caused by alternately increasing and diminishing the resistance of the circuit, or by alternately increasing and diminishing the power of the battery. The internal resistance of a battery is diminished by bringing the voltaic elements nearer together, and increased by placing them farther apart. The reciprocal vibration of the elements of a battery, therefore, occasions an undulatory action in the voltaic current. The external resistance may also be varied. For instance, let mercury or some other liquid form part of a voltaic circuit, then the more deeply the conducting-wire is immersed in the mercury or other liquid, the less resistance does the liquid offer to the passage of the current. Hence, the vibration of the conducting-wire in mercury or other liquid included in the circuit occasions undulations in the current. The vertical vibrations of the elements of a battery in the liquid in which they are immersed produces an undulatory action in the current by alternately increasing and diminishing the power of the battery.

In illustration of the method of creating electrical undulations, I shall show and describe one form of apparatus for producing the effect. I prefer to employ for this purpose an electro-magnet, A, Fig. 5, having a coil upon only one of its legs b. A steel-spring armature, c, is firmly clamped by one extremity to the uncovered leg d of the magnet, and its free end is allowed to project above the pole of the covered leg. The armature c can be set in vibration in a variety of ways, one of which is by wind, and, in vibrating, it produces a musical note of a certain definite pitch.

When the instrument A is placed in a voltaic circuit, $g\,b\,e\,f\,g$, the armature c becomes magnetic, and the polarity of its free end is opposed to that of the magnet underneath. So long as the armature c remains at rest, no effect is produced upon the voltaic current, but the moment it is set in vibration to produce its musical note a powerful inductive action takes place, and electrical undulations traverse the circuit $g\,b\,e\,f\,g$. The vibratory current passing through the coil of the electro-magnet f causes vibration in its armature h when the armatures $c\,h$ of the two instruments A I are normally in unison with one another; but the armature h is unaffected by the passage of the undulatory current when the pitches of the two instruments are different.

A number of instruments may be placed upon a telegraphic circuit, as in Fig. 6. When the armature of any one of the instruments is set in vibration all the other instruments upon the circuit which are in unison with it respond, but those which have normally a different rate of vibration remain silent. Thus, if A, Fig. 6, is set in vibration, the armatures of A^1 and A^2 will vibrate also, but all the others on the circuit will remain still. So if B^1 is caused to emit its musical note the instruments B B^2 respond. They continue sounding so long as the mechanical vibration of B^1 is continued, but become silent with the cessation of its motion. The duration of the sound may be used to indicate the dot or dash of the Morse alphabet, and thus a telegraphic dispatch may be indicated by alternately interrupting and renewing the sound.

When two or more instruments of different pitch are simultaneously caused to vibrate, all the instruments of corresponding pitches upon the circuit are set in vibration, each responding to that one only of the transmitting instruments with which it is in unison. Thus the signals of A, Fig. 6, are repeated by A^1 and A^2, but by no other instrument upon the circuit; the signals of B^2 by B and B^1; and the signals of C^1 by C and C^2--whether A, B^2, and C^1 are successively or simultaneously caused to vibrate. Hence by these instruments two or more telegraphic signals or messages may be sent simultaneously over the same circuit without interfering with one another.

I desire here to remark that there are many other uses to which these instruments may be put, such as the simultaneous transmission of musical notes, differing in loudness as well as in pitch, and the telegraphic transmission of noises or sounds of any kind.

When the armature c, Fig. 5, is set in vibration the armature h responds not only in pitch, but in loudness. Thus, when c vibrates with little amplitude, a very soft musical note proceeds from h; and when c vibrates forcibly the amplitude of the vibration of h is considerably increased, and the resulting sound becomes louder. So, if A and B, Fig. 6, are sounded simultaneously, (A loudly and B softly,) the instruments A^1 and A^2 repeat loudly the signals of A, and $B^1\,B^2$ repeat softly those of B.

One of the ways in which the armature *c*, Fig. 5, may be set in vibration has been stated above to be by wind. Another mode is shown in Fig. 7, whereby motion can be imparted to the armature by the human voice or by means of a musical instrument.

The armature *c*, Fig. 7, is fastened loosely by one extremity to the uncovered leg *d* of the electro-magnet *b*, and its other extremity is attached to the center of a stretched membrane, *a*. A cone, A, is need to converge sound vibrations upon the membrane. When a sound is uttered in the cone the membrane *a* is set in vibration, the armature *c* is forced to partake of the motion, and thus electrical undulations are created upon the circuit E *b e f g*. These undulations are similar in form to the air vibrations caused by the sound--that is, they are represented graphically by similar curves.

The undulatory current passing through the electro-magnet *f* influences its armature *h* to copy the motion of the armature *c*. A similar sound to that uttered into A is then heard to proceed from L.

In this specification the three words "oscillation," "vibration," and "undulation," are used synonymously, and in contradistinction to the terms "intermittent" and "pulsatory." By the terms "body capable of inductive action," I mean a body which, when in motion, produces dynamical electricity. I include in the category of bodies capable of inductive action--brass, copper, and other metals, as well as iron and steel.

Having described my invention, what I claim, and desire to secure by Letters Patent is as follows:

1. A system of telegraphy in which the receiver is set in vibration by the employment of undulatory currents of electricity, substantially as set forth.
2. The combination, substantially as set forth, of a permanent magnet or other body capable of inductive action, with a closed circuit, so that the vibration of the one shall occasion electrical undulations in the other, or in itself, and this I claim, whether the permanent magnet be set in vibration in the neighborhood of the conducting-wire forming the circuit, or whether the conducting-wire be set in vibration in the neighborhood of the permanent magnet, or whether the conducting-wire and the permanent magnet both simultaneously be set in vibration in each other's neighborhood.
3. The method of producing undulations in a continuous voltaic current by the vibration or motion of bodies capable of inducing action, or by the vibration or motion of the conducting-wire itself, in the neighborhood of such bodies, as set forth.
4. The method of producing undulations in a continuous voltaic circuit by gradually increasing and diminishing the resistance of the circuit, or by gradually increasing and diminishing the power of the battery, as set forth.

5. The method of, and apparatus for, transmitting vocal or other sounds telegraphically, as herein described, by causing electrical undulations, similar in form to the vibrations of the air accompanying the said vocal or other sound, substantially as set forth.

In testimony whereof I have hereunto signed my name this 20th day of January, A.D. 1876

ALEX. GRAHAM BELL.

Witnesses:
    THOMAS E. BARRY,
    P. D. RICHARDS.

A. G. BELL
TELEGRAPHY

No. 174,465   Patented March 7, 1876.

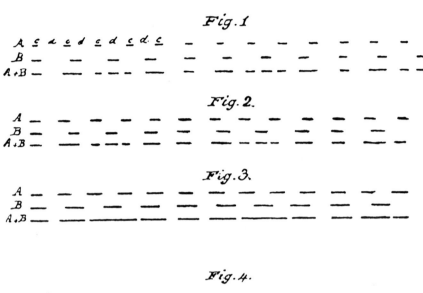

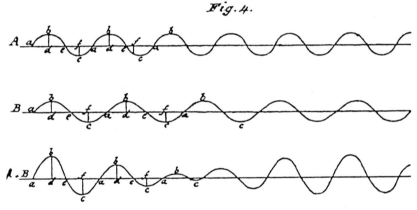

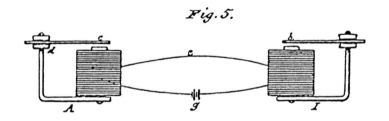

A. G. BELL
TELEGRAPHY

No. 174,465 — Patented March 7, 1876.

Fig 6.

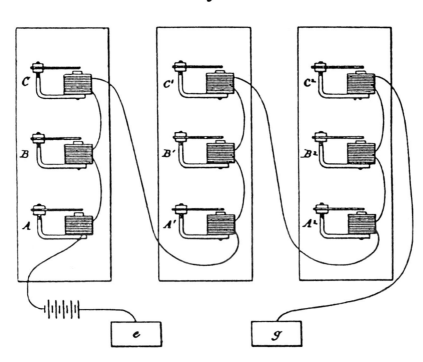

Fig. 7

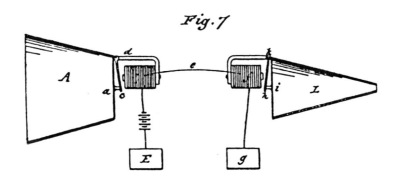

LEE DE FOREST

LEE DE FOREST was born in Council Bluffs, Iowa in 1873 and died in Hollywood, California in 1961. He was most famous for his invention of the "Audion", the first triode vacuum tube in 1907. The triode vacuum tube has since been made obsolete by semiconductor devices, but it was the triode that allowed the start of the modern electronics industry. He also produced over 300 other patented inventions, many of these in the areas of telegraphy, radio, talking pictures, fax transmission, television, radiotherapy, and radar.

Following is an article Dr. Lee de Forest wrote for the Electrical Experimenter, March 1919.

"HOW I INVENTED THE AUDION"

The first conception of a detector of Hertzian waves which should employ the medium of heated electrodes or heated gas came to me as follows:

In the summer of 1900 I was experimenting on a new type of electrolytic detector doing this work at night in my room in Chicago. The receiving apparatus was placed on a table

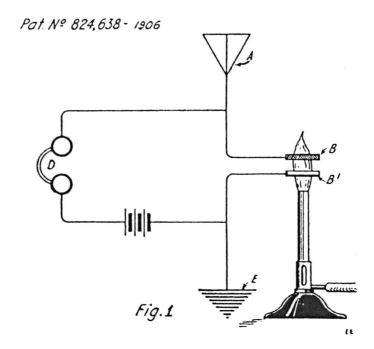

The First State in the Development of the Audion by Dr. Lee de Forest. The Aerial and Ground Were Connected to Two Electrodes Placed in the Flame of a Bunsen Burner, with Successful Results.

beneath a Welsbach gas burner. A spark coil which I was using as my source of oscillations was located in a closet about ten feet distant. One night I noticed that whenever I closed the switch of the spark coil by means of a string running across the floor from my table to the coil there was a decided change in the illumination from the Welsbach burner. The light from the gas mantel increased very perceptibly and resumed its normal low brilliance as soon as the sparking ceased. This phenomena continued and imprest itself strongly upon my attention.

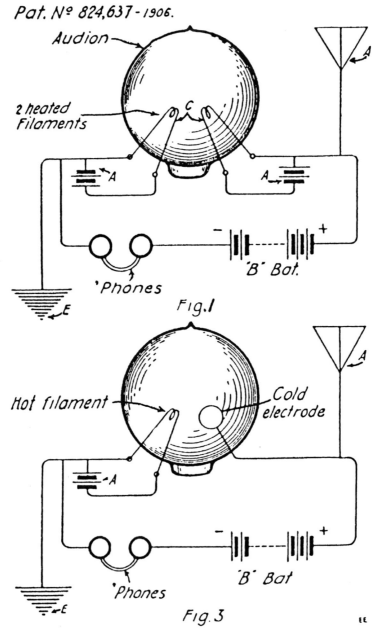

Here the Placing of the Two Heated Electrodes in a Vacuum Bulb Is Carried Out, as Well as the Final Adoption of the Cold "Wing." This Device Works Equally Well with Both Electrodes Hot, Thus Disproving Any "Rectification" Theory.

My first thought was that I had discovered a new form of detector of Hertzian waves, of extraordinary sensitiveness, and was, naturally, much enthused, as any young investigator would have been under similar circumstances. But upon closer investigation of this novel phenomena I found that when the door of the closet was closed, or almost closed, the effect of the spark upon the gas burner ceased! This proved conclusively that I was dealing with sound waves coming upon a sensitive flame and not with electrical waves.

The delusion lasted, however, long enough to force upon my mind the conviction that heated gas molecules were sensitive to high frequency electrical operations, and I determined to investigate further at my first opportunity and actually discover evidence to substantiate my theory. I was unable to do this until the fall of 1902 or '03 when I returned to my gas mantel experiment. I first attempted to investigate the new detector phenomena by using two needles of steel, or platinum, placed close together in the incandescent Welsbach mantel. These two

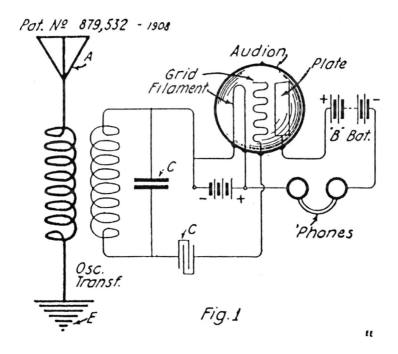

Later the "Third Electrode" Resolved Itself Into a Grid, Which Was Placed Between the Wing and Filament Where It Could Act with the Highest Efficiency. In This Form It Has remained Practically the Same to the Present Time.

needles were connected to a dry battery and telephone receiver. I was, however, unable to obtain any appreciable current between the two electrodes in the mantel. I then investigated the flame of a Bunsen burner and soon found a point in the outside envelope of the flame where an appreciable current did pass between the two electrodes, making a soft fluttering sound in the telephone receiver. (See Fig. 1, Patent No. 824,638, issued in 1906.) Then connecting one electrode to an antenna and the other to the earth, I heard for the first time signals in the telephone receiver; signals which represented clearly the sound of the transmitting spark. Here at last was actually demonstrated my earnest belief in the existence of the this new detector principle. My next step was to enrich the gas flame by putting a lump of potassium or sodium salt in the flame directly below the two platinum electrodes. This increased ionization caused increased flow of battery current, and a corresponding increase in sensitiveness of the new detector. I did considerable work then with various types of Bunsen burner arrangements for permanently enriching the gas flame, etc., and set up a laboratory type of flame detector which was actually used in 1903 for receiving signals from ships down the Harbor of New York.

The inconvenience of supplying a source of gas for the new detector was, of course, obvious, and I sought for other means of obtaining the necessary heated gas and heated electrodes. The electric arc first suggested itself. I anticipated that while this arc would be a

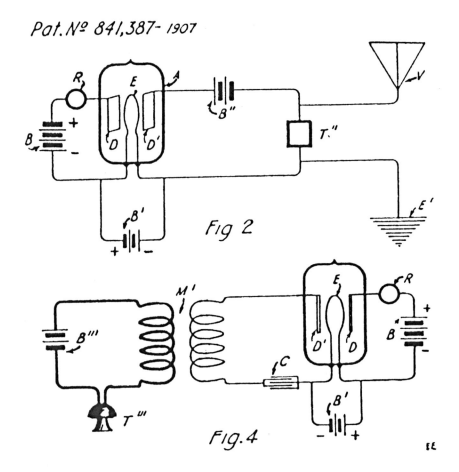

Here We Find the "Third Electrode" Placed Inside the Bulb, Where It Logically Belonged, as Dr. de Forest Points Out. Fig. 4 Shows the First Use of the "Stopping" Condenser.

detector, it would be exceedingly irregular and noisy in the telephone receiver. This was found to be the fact. The battery fed the arc thru the primary of the transformer, in the secondary of which was connected the telephone receiver, and, altho at times the looked for response to the electric waves was thus obtained, the noise in the telephone receiver from the arc was so deafening that the idea was abandoned.

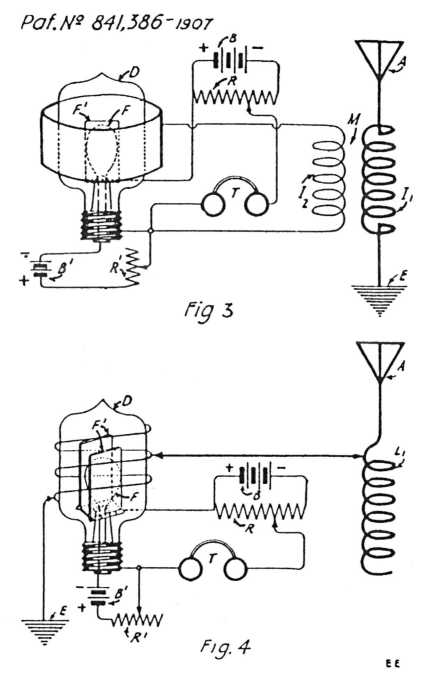

The First "Three Electrode" Audion Had a Piece of Tinfoil Wrapt Around the Outside of the Bulb--a Principle Frequently "Rediscovered", as Was the de Forest Electromagnetic or Coil Electrode, Fig. 4.

The next plan which suggested itself to me was to use incandescent filaments in an enclosed chamber. This arrangement as well as the gas detector, was illustrated in my patent No. 979,275, which bears the first date of November 4, 1904. This patent application was not filed until the following February. The drawings, Figs. 2 and 6, of this patent show the incandescent or glow members both in the air, and sealed within a closed chamber.

It will be noted now that I approached the general problem of this new type of detector from an entirely different angle from that commonly supposed to-day. In the first place I always employed a battery, and this original battery was what is now universally called the "B battery." My source of electric current for heating purposes was second, and secondary so that the vociferous contention of the advocates of the Fleming valve that the audion evolved from the Fleming valve, and was originally the Fleming valve with the "B battery" added as an afterthought, is entirely untrue.

But Fig. 5 of this patent, showing the inclosed filament in a vacuum bulb, is interesting from another consideration because it contains the *first embryonic germ* of the later "grid" or *third electrode*. It was realized from the very first that a certain proportion of the high frequency energy from the antenna could be lost thru the bypath circuit supplied by the battery and telephone receiver. In order to prevent this the arrangement shown in Fig. 5 was tried out, wherein the local and high frequency circuits are kept separate. In this arrangement, as actually tried in the gas flame, no actual advantage was observed, because the high frequency electrodes were necessarily some slight distance from the path conducting the direct current, and hence the effect of the high frequency currents upon the ions carrying the local current was weakened.

During the years of 1904 and '05 my duties kept me almost continually away from my laboratory, in traveling about the country directing the installation of numerous radio stations; consequently there was little opportunity for prosecuting this development work, and carrying out the designs and sketches which I made from time to time. In fact, it was not until 1905 that a lamp manufacturer was found to undertake the construction of the various experimental forms of lamps which I had designed as a successor to the flame or arc detector. I was familiar in 1905 with Prof. Fleming's work on the subject of the "Edison Effect" as utilized for the rectification of high frequency oscillations, or currents. This was interesting to me only as an evidence of growing activity along lines similar to those of my new detector. I was familiar, of course, with the phenomena originally discovered by Edison, and investigated and developed by Howell, Fleming, Wehnelt, et al; but from the very first of my work with radio detectors I always had in mind a *relay* in the true sense of the word, *not a rectifier*; in other words, a detector in which the energy of the audible signal was supplied from a local battery. This local energy being merely controlled or released by the incoming signal. It had always been obvious to me that such a device should be more efficient that any form of rectifier where the signal indication is effected *only* by the energy actually received thru the antenna.

Fig. 1, Patent No. 824,637, is an obviously practical development of Fig. 5 of my first patent, No. 979,275. In this figure two incandescent filaments are sealed in the glass bulb, and each lighted from its own "A" battery. Here, as always, the original separate "B" battery is shown.

This figure is interesting to many, who until recently, if even now, have never heard that an audion of this type works equally well whether both electrodes are incandescent, or whether one is incandescent and the other cold. It might be well to point out now in this connection that during the patent trial--the Fleming valve vs. the Audion--a demonstration was made before the court where all three electrodes were in the form of filaments, where each could be heated to incandescence by a separate battery. The demonstration showed that when this device was connected up as an audion that either outside filament could be used as the "plate" electrode indifferently, and the signals were of the same intensity whether two electrodes were cold or all three hot. The purpose of this test was to demonstrate beyond all cavil the falseness of the assumption that *rectification* plays in any way an important or essential part in the operation of the three electrode audion.

Fig. 3 of the above mentioned patent No. 824,637 shows the next obvious step in the evolution of the audion, i.e., doing away with the unnecessary battery for heating one of the electrodes. It was obvious, of course, that so long as the "B" battery was properly connected so that its positive pole led to the cold electrode, there was no necessity or advantage in heating this second electrode. The audion in the form--one hot and one cold electrode, was used for some months, and shows a sensitiveness as a detector, superior to that of the electrolytic detector, and far superior to the Fleming's valve rectifier. We had, thanks to the "B" battery which was invariably employed, a genuine *relay* or *trigger action* of the high frequency oscillations upon the normal current-carrying ions, or electrons, passing between the two electrodes.

As everyone familiar with incandescent lamp or X-ray bulb phenomena knows, the proportions of current passing between two electrodes therein (one or both being hot) carried by ions or carried by thermions, depends chiefly upon the degree of exhaustion of the bulb. The gradual preponderance of thermionic conduction over ionic conduction has been gradually increasing as the art has progrest with improvements in pumps, knowledge of the fine points of exhaustion, etc. Thus it has always been impossible to lay a finger upon a certain date or upon any audion type of device, and say, "This marks the distinction between an ionic, or gaseous detector, and a thermionic detector." It is in fact impossible to prove or even to-day when audions or oscillations are exhausted to the highest degree that the conductivity is *entirely electronic*.

In the spring and summer of 1906 I had opportunity to spend a good deal of time on the audion problem, and was always seeking to improve its efficiency. Keeping in mind then the disadvantage of directly connecting the high frequency circuit with the local circuits, and harking back to the four electrode gas-flame detector above mentioned, I sought to keep one electrode of the high frequency circuit distinct from the two electrodes of the local current.

Obviously the most simple experiment was to *wrap a piece of tinfoil around the outside of the glass bulb and connect this to one end of the secondary transformer of my receiver*. The other end of the transformer was connected to the filament, it being obviously unnecessary to employ four electrodes to effect the end desired. Exactly this arrangement, with the third electrode around the outside of the bulb, is shown in Fig. 3 of patent No. 841,386 which was filed in August, 1906. It will probably be recalled that this outside electrode has been very recently "re-discovered" with considerable eclat and acclaim! I also at this time wrapt a coil of wire around the bulb, connecting one end of this to the antenna and the other to the ground, seeking thus to effect the ionic conductivity between the inner electrodes by electro-magnetic influence from the high frequency oscillations passing around this helix. See Fig. 4, patent No. 841,386.

The arrangement of the external tinfoil belt may therefore be called the *parent of the third electrode*. It showed a decided improvement in the sensitiveness of the detector, as I had anticipated. I recognized that by this arrangement I had in effect a condenser between the filament connection and a hypothetical third electrode, which consisted of the conducting layer of gas located on the interior walls of the bulb, the other arm of this condenser being the tinfoil belt outside the glass. I recognized also that this was a very inefficient and indirect way of impressing the effect of the high frequency oscillations upon the conducting medium between the filament and plate. The third electrode should therefore be placed inside the bulb. I immediately instructed McCandless & Company to make such a bulb. The first type of this third electrode was in the form of a plate, located on the opposite side of the filament from the "B" battery plate. This arrangement showed the increased efficiency and sensitiveness anticipated. It is shown in Figs. 2 and 4 of patent No. 841,387, filed October, 1906. This is the audion amplifier and telephone relay patent. Fig. 2 of this patent is interesting as showing also for the first time a third battery ("B") in the external circuit between the third electrode and the filament. T in this figure represents the high frequency transformer.

In Fig. 4, where this battery is omitted, is shown for the first time the grid stopping condenser C. In studying this type of three electrode bulb, I recognized that the third electrode was not yet in its most efficient position. It should be placed directly in the path of the ionic or thermionic stream, passing from filament to plate, where the high frequency electric charges imprest on the electrode could best affect this stream. But if placed directly between the two electrodes, a solid plate, of course, would constitute practically a complete barrier. Hence I devised the *grid* or *perforated screen structure*. In fact, the first audion where the third electrode was placed between the filament and plate utilized the wire bent in grid form which is familiar to every amateur or user of the audion prior to 1914.

This type of third electrode so located was so marked an improvement over the preceding three-electrode bulb, that shortly thereafter a patent was applied for on it. This was issued in February, 1908, No. 879,532. See Fig. 1 where the complete receiving arrangement and the grid audion is clearly shown.

The audion remained in this form for six years. During that time its merits became gradually recognized in Europe as well as here, and it was not long before the little stranger was, like its predecessor, the two electrode brother with "B" battery, adopted into the Marconi family, and like it predecessor re-christened the "Fleming Valve." As soon as the audion amplifier had been developed for long distance telephone service by the engineers of the Western Electric Company, and installed on most other long distance lines of the A. T. & T. Co., we find certain English publications adopting it also into the Fleming valve family; and now after the three-electrode device has demonstrated its utility as a radio transmitter of absolutely constant undamped waves and made possible transoceanic telephony, we learn that this brother which I first named the Oscillion, is also the Fleming valve. Not even "junior" or "senior" is used to distinguish one from the other in this rapidly growing Fleming valve family.

The art founded on the three-electrode audion has grown of late years with enormous strides. The great way has produced a tremendous intensity of development for various military purposes and it cannot be disputed that the engineers of the Western Electric Company have taken a foremost position, and much of the present-day efficiency of the detector and amplifier has been due to their efforts, spurred on as they were by the difficult demands and specifications of our Signal Corps officers and engineers. It is estimated that there has been constructed for the U. S. Signal Corps during the war between 200,000 and 300,000 audion amplifier bulbs and at least 50,000 small oscillators. In Great Britain wartime production has probably equaled or exceeded the above; while in France we are informed that during the last two years of the war, the audion production has averaged about *5,000 per day*! The French bulb is particularly interesting as being efficient and suitable in all three uses, detector, amplifier and oscillator. For such purposes, of course, a compromise in efficiency was inevitable, and maximum efficiency in either of these three branches has been somewhat sacrificed.

Considerable discussion has lately arisen as to the first use of the audion as an oscillator or source of alternating current. This matter is now being thrashed out in a multiple interference procedure in the U.S. Patent Office. But the evidence so far indicates that the writer's application of this property of the audion in the spring of 1912 marked the first use of the audion as a generator of undamped electrical currents. In view of recent developments, particularly the highly interesting announcements of President Vail of the A. T. & T. Co., regarding multiplex wire telephony and telegraphy over a single conductor pair, it may be prophesied that the application of the audion as a generator of alternating currents will be fully useful as that of detector and amplifier.

There is, in the writer's opinion, no doubt but that if the development of radio is not now made a Government monopoly, it will not be long before commercial trans-oceanic wireless telephony will be effected. This work, whether the generator be a bank of oscillions or a high frequency alternator, will be made possible only thru the extraordinary amplifying properties of the audion, when used as telephone repeater or relay.

The simplicity of the oscillation transmitter in small sizes, coupled with the extraordinary sensitiveness of the "zero beat" audion detector or amplifier of received high frequency energy, warrants the belief that before long the wireless telephone will be installed on thousands of vessels, supplementing, and in many cases, supplanting the wireless telegraph. In addition there is an enormous number of small vessels where a wireless telephone installation is more feasible.

As indicative of the growth of the Audion Art, the number of patents issued on various devices and circuits dependent thereon, gives a pretty fair key: Up to 1912 there had been issued about 20 patents, all filed subsequent to 1904. To-day there are over 100 United States patents on the Audion Art, and the number is very rapidly growing. Regardless of what name may be applied to the device patented, practically everyone of these patents since 1906 shows the three electrode bulb. They may all therefore be properly described as the outgrowth of the ideas first shown in the audion amplifier--patent No. 841,387.

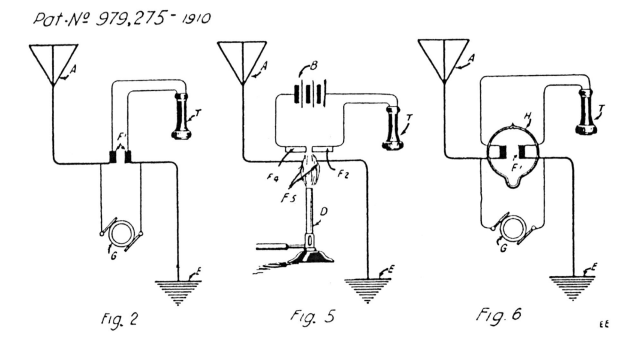

Altho This Patent Bears the Date of 1910, It Was Originally Filed in 1904, and Is One of the Basic Patents in All Audion Development. It Shows the First Use of Incandescent Filaments in an Enclosed Chamber, and, Moreover, It Includes at Fig. 5 the First Evidence of the Third Electrode or "Grid" Conception. Here, as Will Be Evidence of the Third Electrode or "Grid" Conception. Here as Will Be Evident, Is the First Plan for Separating the Antenna Circuit from the Local Circuit.

THOMAS ALVA EDISON was born in Milan, Ohio in 1847 and died in West Orange, New Jersey in 1931. America's favorite workaholic, he is most famous for his invention of the incandescent electric light bulb in 1879. He produced 1,033 patents and gave a major boost to the modern electrical industry. Edison was practically self-taught, having spent only three months in the public schools. He formed the Edison General Electric Company which later merged into the present General Electric Company.

Following sections contain some of the thoughts that Edison expressed between 1914 and 1927. These can be found reprinted in the book by Dagobert D. Runes (editor), <u>The Diary and Sundry Observations of Thomas Alva Edison</u>, (New York: Philosophical Library, 1948) p.161-81.

"MAN AND MACHINE".....dated 1914

I DO NOT BELIEVE the Government should do anything but regulate the activities of its people, give them a free swing, and see that every man is protected in that which he produces. A department of inventions is not wanted. What is wanted is that the methods of court procedure be changed and the courts realize that the man who makes inventions, by the very nature of things, cannot be a business man, familiar with its merciless code; and they should take this into consideration and protect him.

Economic questions involve thousands of complicated factors which contribute to a certain result. It takes a lot of brain power and a lot of scientific data to solve these questions. In the first place, they ought to be studied scientifically, the same way we go about discovering the so-called secrets of nature.

When I want to discover something, I begin by reading up everything that has been done along that line in the past--that's what all these books in the library are for. I see what has been accomplished at great labor and expense in the past. I gather the data of many thousands of experiments as a starting point, and then I make thousands more. On this money question we ought to go back several thousand years before the Roman era and find out all about the financial systems and their results for that time to this. Then we would know something to build upon. Take the tariff question. An item is put in a bill and it is expected to produce tremendous results. The actual result is just nothing. But another item, small and unconsidered, produces enormous changes in the national economy. What do the legislators know about that?

Herbert Spencer had the right idea. He took thirty-two acts of Parliament and had them traced down and found that twenty-nine produced exactly the contrary effect to the effect intended. Spencer had the right scientific idea of investigating economics. He hired thirty clerks to run down those laws and see what their results were.

There are plenty of wrong things in our society. Everything is for show; the newspapers make a show of everything. Things are wrong at the top and at the bottom. Between the two they are fairly tolerable. There isn't too much happiness floating around, and the man who gets nearest to his rightful share of it has a character, a little bungalow in the country, and a family. What does the very rich man get? He's always scheming, always suspicious of the men around him. His money is mostly out, invested. Yes, he lives in a fine house, rides in an automobile, and he eats three meals a day when he feels able to. I defy any one to prove that he gets much out of life. Money doesn't make a man happy and it doesn't make a man a good companion.

Things are wrong enough, and to right them we need two remedies. One is to develop the convolutions in man's brain, those coils inside with which he does his thinking. We have gradually developed what we have in there, and if we could develop about two convolutions more we would be able to grasp and solve our social problems. The other remedy is education. Education of the right sort in early childhood. You can't do anything with a grown man. You can't do anything or predict anything about a woman, either, because she is all instinct and emotion. But take a child of four years old and its mind is plastic, and whatever you put in there will always stay. Teach a child of four that the moon is made of green cheese, and tho you give him a thorough scientific education afterward there will always be, at the bottom of his mind, a feeling that the moon is somehow possibly made of green cheese. See how religious beliefs implanted in childhood stay with the adult in spite of everything. It is necessary to take them young and to teach morality and character, to fix ideas in those plastic minds so that it will be impossible for them to think wrong or do wrong. What we want to do in this world is to eradicate the crooks, high and low, and to do that we must begin early and prevent them from going crooked at the start.

Yes, I know the Socialist viewpoint. I guess the reason for their belief is that they see so much fraud everywhere; they get the seamy side of everything. It's a recommendation, of course, for Socialism that there are so many of the intellectual class who sympathize with it or believe in it. But they'll have to improve their ideas to make them practical. So far Russia is the most socialist country and everything there is like a machine and nobody likes it. They have it in the factories, where, as I saw it in a comic paper over there, they prescribe how many steps to the right and left a workingman takes at the noon hour in going from the factory door to the eating place. They have it in the schools, forcing all kinds of dry stuff into the heads of school children. Learning ought to be made easy and pleasant. It can be done with the aid of moving pictures. I could tell any one a great deal about a dynamo and it would be hard for him to understand; but I could show everything in a few pictures so that a child would understand--and would never forget.

Now, the Socialists, if they amount to anything, must improve their program--or what is generally accepted as their program. They can't hope to reduce all mankind to a dead level. They can't figure to abolish capital, which is the accumulated results of labor, mental and physical, of all the ages, and is called wealth, wealth of all the ages. They can't ignore the men who do the thinking and guiding, the great executive minds to whom society owes most of what it has. Two men start two factories, with the same resources, on opposite sides of the street. One goes bankrupt, the other succeeds. Are those men equal? Or here is a man who goes into a shipyard and without increasing the hours of labor or making any one work harder, manages it so that three ships instead of two are built in a year. This he has done without calling for any more exertion on the part of the men and without increasing their number. Didn't he create extra value and isn't he entitled to extra reward? Such men are not in the class of parasites or market manipulators or stock jugglers. Socialism, if it ever arrives, must provide unlimited incentive for its executive minds and its creators. Unlimited incentive. The motive that I have for inventing is, I guess, like the motive of the billiard player, who always wants to do a little better--to add to his record. Under present conditions I use the reasonable profit which I derive from one invention to make experiments looking toward another invention. If socialism gave me the means to continue inventing, I would invent; but if it failed to do so, or began to tie me down, I would quit.

Machinery has changed things in our society and it will change them a great deal more. The man and the machine act and interact. The time is coming when the machine will do all the work and man will just set it to work. We will feed the raw material in one end and will see our shoes, clothes and everything else we need come out of the other end. The general use of such automatic machinery will be forced by the tactics of radical labor, and at first the working people will suffer, but in the end they will be benefited.

"THE INVENTOR'S LOT".....dated 1914

THE INVENTOR tries to meet the demand of a crazy civilization. Society is never prepared to receive any invention. Every new thing is resisted, and it takes years for the inventor to get people to listen to him and years more before it can be introduced, and when it is introduced our beautiful laws and court procedure are used by predatory commercialism to ruin the inventor. They don't leave him even enough to start a new invention.

"THEY WON'T THINK".....dated 1921

EVERY MAN has some forte, something he can do better than he can do anything else. Many men, however, never find the job they are best suited for. And often this is because they do not think enough. Too many men drift lazily into any job, suited or unsuited for them; and when they don't get along well they blame everybody and everything but themselves.

Grouches are nearly always pinheads, small men who have never made any effort to improve their mental capacity.

The brain can be developed just the same as the muscles can be developed, if one will only take the pains to train the mind to think.

Why do so many men never amount to anything? Because they don't think.

I am going to have a sign put up all over my plant, reading "There is no expedient to which a man will not resort to avoid the real labor of thinking."

That is true. There is hardly a day that I do not discover how painfully true it is.

What progress individuals could make, and what progress the world would make, if thinking were given proper consideration! It seems to me that not one man in a thousand appreciates what can be accomplished by training the mind to think.

It is because they do not use their thinking powers that so many people have never developed a creditable mentality. The brain that isn't used rusts. The brain that is used responds. The brain is exactly like any other part of the body: it can be strengthened by proper exercise, by proper use. Put your arm in a sling and keep it there for a considerable length of time, and, when you take it out, you find that you can't use it. In the same way, the brain that isn't used suffers atrophy.

By developing your thinking powers you expand the capacity of your brain and attain new abilities. For example, the average person's brain does not observe a thousandth part of what the eye observes. The average brain simply fails to register the things which come before the eye. It is almost incredible how poor our powers of observation--genuine observation--are.

Let me give an illustration: When we first started the incandescent lighting system we had a lamp factory at the bottom of a hill, at Menlo Park. It was a very busy time for us all. Seventy-five of us worked twenty hours every day and slept only four hours--and thrived on it.

I fed them all, and I had a man play an organ all the time we were at work. One midnight, while at lunch, a matter came up which caused me to refer to a cherry tree beside the hill leading from the main works to the lamp factory. Nobody seemed to know anything about the location of the cherry tree. This made me conduct a little investigation, and I found that although twenty-seven of these men had used this path every day for six months not one of them had ever noticed the tree.

The eye sees a great many things, but the average brain records very few of them. Indeed, nobody has the slightest conception of how little the brain 'sees' unless it has been highly trained. I remember dropping in to see a man whose duty was to watch the working of a hundred machines on a table. I asked him if everything was all right.

Yes, everything is all right, he said.

But, I had already noticed that two of the machines had stopped. I drew his attention to them, and he was mortified. He confessed that, although his sole duty was to watch and see that every machine was working, he had not noticed that these two had stopped. I could hide myself off and keep busy at thinking forever. I don't need anybody to amuse me. It is the same way with my friends John Burroughs, the naturalist, and Henry Ford, who is a natural-born mechanic. We can derive the most satisfying kind of joy from thinking and thinking and thinking.

The man who doesn't make up his mind to cultivate the habit of thinking misses the greatest pleasure in life. He not only misses the greatest pleasure, but he cannot make the most of himself. All progress, all success, springs from thinking.

Of course, even the most concentrated thinking cannot solve every new problem that the brain can conceive. It usually takes me from five to seven years to perfect a thing. Some things I have been working on for twenty-five years--and some of them are still unsolved. My average would be about seven years. The incandescent light was the hardest one of all; it took many years not only of concentrated thought but also of world-wide research. The storage battery took eight years. It took even longer to perfect the phonograph.

Which do I consider my greatest invention? Well, my reply to that would be that I like the phonograph best. Doubtless this is because I love music. And then it has brought so much joy into millions of homes all over the country, and, indeed, all over the world. Music is so helpful to the human mind that it is naturally a source of satisfaction to me that I have helped in some way to make the very finest music available to millions who could not afford to pay the price and time necessary to hear the greatest artists sing and play.

Many inventions are not suitable for the people at large because of their carelessness. Before a thing can be marketed to the masses, it must be made practically fool-proof. Its operation must be made extremely simple. That is one reason, I think, why the phonograph has been so universally adopted. Even a child can operate it.

Another reason, is that people are far more willing to pay for being amused than for anything else.

One great trouble with the world to-day is that people wander from place to place, and are never satisfied with anything. They are shiftless and thoughtless. They revolt at buckling down and doing hard work and hard thinking. They refuse to take the time and the trouble to lay solid foundations. They are too superficial, too flighty, too easily bored. They fail to adopt the right spirit toward their life work, and consequently fail to enjoy the satisfaction and the happiness which comes from doing a job, no matter what it is, absolutely in the best way within their power. Failing to find the joy which they should find in accomplishing something, they turn to every imaginable variety of amusement. Instead of learning to drink in joy through their minds, they try to find it, without effort, through their eyes and their ears--and sometimes their stomachs.

It is all because they won't think, won't think!

"THEY DO WHAT THEY LIKE TO DO".....dated 1921

PEOPLE WILL NOT only do what they like to do--they overdo it 100 per cent. Most people overeat 100 per cent, and oversleep 100 per cent, because they like it. That extra 100 per cent makes them unhealthy and inefficient. The person who sleeps eight or ten hours a night is never fully asleep and never fully awake--they have only different degrees of doze through the twenty-four hours. Most people seem to think they must eat until they are no longer hungry. Most of their energies are taken up in digesting what they eat. I see what people eat and for myself half as much is enough.

For myself I never found need of more than four or five hours' sleep in the twenty-four. I never dream. It's real sleep. When by chance I have taken more I wake dull and indolent. We are always hearing people talk about "loss of sleep" as a calamity. They better call it loss of time, vitality and opportunities. Just to satisfy my curiosity I have gone through files of the British Medical Journal and could not find a single case reported of anybody being hurt by loss of sleep. Insomnia is different entirely--but some people think they have insomnia if they can sleep only ten hours every night.

Now, I'm not offering advice. That's no use. Nobody takes advice. As I say, people do what they like to do and overdo it 100 per cent, and the same rule applies to the giving of advice that nobody pays any attention to. The world is badly overstocked with unused advice.

"MACHINE AND PROGRESS".....dated 1926

IT HAS BEEN CHARGED abroad and occasionally at home that we of the United States have become a machine-ridden people, that we are developing upon lines too completely mechanical. The very reverse is the truth. We are not mechanical enough. The machine has been the human being's most effective means of escape from bondage. Too many people, even now, remain bond-slaves to laborious hand-processes. Not through fewer, but through more machines, not through simpler, but through more complex machines, will men find avenues that lead into lives of greater opportunity and happiness.

We substitute motors for muscles in a thousand new ways. A human brain is greatly hampered in its usefulness if it has only two hands of a man to do its bidding. There are machines each of which can do the work of a multitude of hands, when directed by one brain. That is efficiency.

Anything which tends to slow work down is waste. Every effort should be made to speed work up. Increased production means enlarged lives for mankind. Human hands alone can do no more than they did long ago by way of fast production. Only machines, not nerves and muscles, can increase men's output. We have scarcely seen the start of the mechanical age, and after it is under way we shall discover that it is also a mental age as never has been known before. One of the reasons it will be notably mental will be that it will be notably mechanical. It requires a surprising amount of complexity to displace the mechanical effort of the man. The difference between the automatic and the semi-automatic machine is very great. Its significance in industry is immense. But once the fully automatic has been achieved, the output and quality of the product will be greatly increased. All fully automatics, on account of their very complexity, require attendants of mental capacity greatly increased over that of men who are merely parts of semi-automatics.

There could be no greater waste than keeping good brains at work directing hands of the bodies they control in the hand-execution of mechanical tasks because of the mere failure to invent and develop machines to execute those tasks better and faster than hand work can execute them under good brain direction. Man will progress in intellectual things according to his release from the mere motor-tasks.

The history of slavery is full of illustrations of the value of machinery. Slavery, the use of men as beasts of burden and as motors, was mental bondage for the men who thought they benefited by it, as well as physical bondage for the men they held in thrall. While slave labor was available, the brains of men in general were not stimulated to the creation of machinery. This was more disastrous in its general effects than was realized by the majority, even of those opposed to slavery. I meant that human beings all along the line, not only the enslaved but the enslavers, could not be released by machinery for efforts better and more elevating than those to which they had been habituated in the past. Progress of mind became impossible.

That is the reason why I call machinery the greatest of emancipators. I will go farther and say that human slavery will not have been fully abolished until every task now accomplished by human hands is turned out by some machine, if it can be done as well or better by a machine. Why chain a man, thus wasting him, to laborious work which a machine could do? All men cannot walk out of the shadow into the light until all men understand the foolishness of such procedure.

The shoe factory of to-day requires better employees than were required by the old processes of laborious, slow, hand work. Some of the old time cobblers were fine fellows who could think, but they would have thought far more and better if their ignorance of machinery had not shackled them to the awl and hand-hammer-driven peg, to the bristle-tipped waxed-end. These things did nothing then which now are not far better done by our machines. I have said that men's brains are bettered by machinery, if it is of the right sort. That with which we now make boots and shoes develops brains far more fully than work with the old tools did or could. This is proved by the fact that when, by working at machines, men's brains are improved sufficiently, the men who have shown ability to run the first machines are promoted to the operation of those which are now more complicated and run still faster, requiring of their operators increased alertness and mentality.

There is no common-sense in the cry that machine work is monotonous. On the other hand it creates a good product, uniform and universally dependable, which is something hand-work never could do. Machine work robs the product of the ill-effects of man's changing physical and mental conditions. The hand worker's product is uneven. Far too much of it is too bad to enable it to compete successfully with the output of the machine.

Americans use more machinery than anybody else. I am told that American workers each can run six looms of a certain kind, Germans five, Frenchmen five, Englishmen (whose workers have never ceased agitating against machinery) five, --and Chinese one, the quality being the same when the cloth is inspected. If the Chinese should begin suddenly to use machinery extensively, it would be only a matter of time when they could run more machines than they can now. Their indicator numeral would go up in the scale. But at the start, one Chinese could run but one machine. The workers of the so-called "machinized" nations can operate the larger numbers I have stated because working with machines has much increased their mental development.

If we continue to increase machine production and the number of machines engaged at it, the next generation will be far beyond where we are now in its intelligence as well as in the possession of facilities for getting the good things out of life.

One of the most foolish things men say, and one which they often repeat, is that too much substitution of machine-work for hand-work will bring over-production. The idea is complete nonsense. There cannot be over-production of anything which men and women want, and their wants are unlimited except in so far as they are limited by the size of their stomachs. The stomach is the only part of man which can be fully satisfied. The yearning of man's brain for new knowledge and experience and for pleasanter and more comfortable surroundings never can be completely met. It is an appetite which cannot be appeased. Talk of over-production is a bugaboo.

A general benefit ensues inevitably for the increasing use of machinery. Not only do the workers benefit through the development enforced on them by the machines, but, in exact proportion as the machines enable the manufacturers to turn out more and better work, the sale of their manufactures is permitted at a decreased price. If the manufacturer can sell at a decreased price then, automatically, it becomes possible for the man of average income to have more things than theretofore. That man of average income has gained tremendously through the creation of machines. There is no doubt in my mind that in quantity production, so called, lies the greatest hope which now exists to cheer the human race. Quantity production cannot possibly occur without machinery. Therefore no man should rail against machine-power. It is application of good fertilizer to industry.

We use every known device of science, and continually seek for new ones, in our efforts to enlarge the production of our grain fields, our fruit orchards, and our vegetable gardens. What is that but striving to stimulate our plants and trees to quantity production? Are not the productive powers of men as worthy of good fertilizer? Machinery is the influence which enables men to do what stimulated plant life does to increase this year's output as compared with last year's to make certain for next year of more than this year's yields. It is as worthy an ambition to make two pairs of shoes where one was made before, that is, with the same human effort, as it is to devise agricultural means of making two blades of grass grow this year where only one grew last.

When objection to machinery has occurred among the workers it has been as foolish as is the refusal of men to accept any other opportunity for progress. Time was when printers all over New York City, and, indeed, the nation, struck or threatened to strike against typesetting machines, fearing that if they should come into general use fewer printers would be hired and at lower wages. The machine won, of course. And there are far more printers working now than were working then, and wages are higher. The economic status of the printer has much improved. So has his intelligence. So has his self-respect. He does not have to do things with his brains and muscles which a machine without brains can do better and faster. The printers of to-day would strike if you should try to take machines away from them.

The history of the typesetting machine is like that of every other machine which has been introduced to perform work previously done by hand; laboriously, slowly, expensively, and less perfectly. The sewing-machine, for instance, has increased by fifty-fold the employment in the fields which it affects.

Whenever something has compelled us to put in machinery to do work theretofore done by men's hands and muscles while brains have remained comparatively idle, all of us, and especially the men directly involved, have gained. If labor everywhere would strike against the use of men as animals instead of protesting against their use as human beings, it would show superior wisdom. Such a strike would have an unprecedentedly good effect on human life, for in very many of our most important activities the possibilities of machine development as an accessory to human intelligence and productiveness are, even yet, not fully understood.

"THE DESIRE FOR CHANGE".....dated 1927

PERPETUAL YOUTH and virtual immortality on this earth would seem to me to be most undesirable. When the time comes, normal human beings do not desire abnormal extension of the earthly life-period. No dreamer about immortality has crystallized his dreams into a desire for a perpetual extension of such lives as we live here. Enough's enough of any human life as human lives are now. Those normal men who have reached the extreme limit of the human life cycle invariably are indifferent to death. They do not desire extension of the present existence. The group of entities which make up such a normal man's intelligence seek release from, rather than prolongation of, existence in the conditions and environments of this cycle so that they may enter another, whatever it may be. All through life humanity yearns for change, for without change progress is impossible and I am convinced that at the end of that which we call life this subconscious desire for something new is very great, and in many instances influential, no matter how the conscious mind, trained by instinct and long habit to cling to this existence, may struggle to combat it. New scenes, new occupations, new emotions, new successes,--these all normal human beings strive for during this life. When they have had all of these that they can get out of it they must turn for change to whatever may come beyond.

"AGE AND ACHIEVEMENT".....dated 1927

THE MAN who has reached the age of thirty-six has just about achieved readiness to discard the illusions built on the false theories for which wrong instruction and youthful ignorance previously have made him an easy mark. He is just beginning to get down to business. If he is really worth while he has passed through a series of hard knocks by that time. The useful man never leads the easy, sheltered, knockless, unshocked life. At thirty-six he ought to be prepared to deal with realities and after about that period in his life, until he is sixty, he should be able to handle them with a steadily increasing efficiency. Subsequently, if he has not injured his body by excess indulgence in any of the narcotics (and by this term I mean, here, liquor, tobacco, tea, and coffee), and if he has not eaten to excess, he very likely may continue to be achievingly efficient up to his eightieth birthday and in exceptional cases until ninety.

Then the curve turns sharply down. The cycle is approaching the end. At about that age the entities which form that man will be preparing to discard their abode, which is that man, and enter upon a new cycle. Then and not till then men should, must and do begin to step aside. If all men did so at the age of thirty-five the world of times to come would be virtually without achievement and leadership.

Following are the patents of Edison's most famous inventions, the Electric Lamp and the Phonograph.

UNITED STATES PATENT OFFICE

THOMAS A. EDISON, OF MENLO PARK, NEW JERSEY

ELECTRIC LAMP.

SPECIFICATION forming part of Letters Patent No. 223,398, dated January 27, 1880

Application filed November 4, 1879

To all whom it may concern:

Be it known that I, Thomas ALVA Edison, of Menlo Park, in the State of New Jersey, United States of America, have invented an Improvement in Electric Lamps, and in the method of manufacturing the same, (Case No. 186) of which the following is a specification.

The object of this invention is to produce electric lamps giving light by incandescence, which lamps shall have high resistance, so as to allow of the practical subdivision of the electric light.

The invention consists in a light-giving body of carbon wire or sheets coiled or arranged in such manner as to offer great resistance to the passage of the electric current, and at the same time present but a slight surface from which radiation can take place.

The invention further consists in placing such burner of great resistance in a nearly-perfect vacuum, to prevent oxidation and injury to the conductor by the atmosphere. The current is conducted into the vacuum-bulb through platina wires sealed into the glass.

The invention further consists in the method of manufacturing carbon conductors of high resistance, so as to be suitable for giving light by incandescence, and in the manner of securing perfect contact between the metallic conductors or leading-wires and the carbon conductor.

Heretofore light by incandescence has been obtained from rods of carbon of one to four ohms resistance, placed in closed vessels, in which the atmospheric air has been replaced by gases that do not combine chemically with the carbon. The vessel holding the burner has been composed of glass cemented to a metallic base. The connection between the leading-wires and the carbon has been obtained by clamping the carbon to the metal. The leading-wires have always been large, so that their resistance shall be many times less than the burner, and, in general, the attempts of previous persons have been to reduce the resistance of the carbon rod. The disadvantages of following this practice are, that a lamp having but one to four ohms resistance cannot be worked in great numbers in multiple are without the employment of main conductors of enormous dimensions; that, owing to the low resistance of the lamp, the leading-wires must be of large dimensions and good conductors, and a glass globe cannot be kept tight at the place where the wires pass in and are cemented; hence the carbon is consumed, because there must be almost a perfect vacuum to render the carbon stable, especially when such carbon is small in mass and high in electrical resistance.

The use of a gas in the receiver at the atmospheric pressure, although not attacking the carbon, serves to destroy it in time by "air-washing," or the attrition produced by the rapid passage of the air over the slightly-coherent highly-heated surface of the carbon. I have reversed this practice. I have discovered that even a cotton thread properly carbonized and placed in a sealed glass bulb exhausted to one-millionth of an atmosphere offers from one hundred to five hundred ohms resistance to the passage of the current, and that it is absolutely stable at very high temperatures; that if the thread be coiled as a spiral and carbonized, or if any fibrous vegetable substance which will leave a carbon residue after heating in a closed chamber be so coiled, as much as two thousand ohms resistance may be obtained without presenting a radiating-surface greater than three-sixteenths of an inch; that if such fibrous material be rubbed with a plastic composed of lamp-black and tar, its resistance may be made high or low, according to the amount of lamp-black placed upon it; that carbon filaments may be made by a combination of tar and lampblack, the latter being previously ignited in a closed crucible for several hours and afterward moistened and kneaded until it assumes the consistency of thick putty. Small pieces of this material may be rolled out in the form of wire as small as seven one-thousandths of a inch in diameter and over a foot in length, and the same may be coated with a non-conducting non-carbonizing substance and wound on a bobbin, or as a spiral, and the tar carbonized in a closed chamber by subjecting it to high heat, the spiral after carbonization retaining its form.

All these forms are fragile and cannot be clamped to the leading wires with sufficient force to insure good contact and prevent heating. I have discovered that if platinum wires are used and the plastic lamp-black and tar material be molded around it in the act of carbonization there is an intimate union by combination and by pressure between the carbon and platina, and nearly perfect contact is obtained without the necessity of clamps; hence the burner and the leading-wires are connected to the carbon ready to be placed in the vacuum-bulb.

When fibrous material is used the plastic lamp-black and tar are used to secure it to the platina before carbonizing.

By using the carbon wire of such high resistance I am enabled to use fine platinum wires for leading-wires, as they will have a small resistance compared to the burner, and hence will not heat and crack the sealed vacuum-bulb. Platina can only be used, as its expansion is nearly the same as that of glass.

By using a considerable length of carbon wire and coiling it the exterior, which is only a small portion of its entire surface, will form the principal radiating surface; hence I am able to raise the specific heat of the whole of the carbon, and thus prevent the rapid reception and disappearance of the light, which on a plain wire is prejudicial, as it shows the least unsteadiness of the current by the flickering of the light; but if the current is steady the defect does not show.

I have carbonized and used cotton and linen thread, wood splints, papers coiled in various ways, also lamp-black, plumbago, and carbon in various forms, mixed with tar and kneaded so that the same may be rolled out into wires of various lengths and diameters. Each wire, however, is to be uniform in size throughout.

If the carbon thread is liable to be distorted during carbonization it is to be coiled between a helix of copper wire. The ends of the carbon or filament are secured to the platina leading-wires by plastic carbonizable material, and the whole placed in the carbonizing-chamber. The copper, which has served to prevent distortion of the carbon thread, is afterward eaten away by nitric acid, and the spiral soaked in water, and then dried and placed on the glass holder, and a glass bulb blown over the whole, with a leading-tube for exhaustion by a mercury-pump. This tube, when a high vacuum has been reached, is hermetically sealed.

With substances which are not greatly distorted in carbonizing, they may be coated with a non-conducting non-carbonizable substance, which allows one coil or turn of the carbon to rest upon and be supported by the other.

In the drawings, Figure 1 shows the lamp sectionally. a is the carbon spiral or thread. $c\ c'$ are the thickened ends of the spiral, formed of the plastic compound of lamp-black and tar. $d\ d'$ are the platina wires. $h\ h$ are the clamps, which serve to connect the platina wires, cemented in the carbon, with the leading-wires $x\ x$, sealed in the glass vacuum-bulb. $e\ e$ are copper wires, connected just outside the bulb to the wires x x. m is the tube (shown by dotted lines) leading to the vacuum-pump, which, after exhaustion, is hermetically sealed and the surplus removed.

Fig. 2 represents the plastic material before being wound into a spiral.

Fig. 3 shows the spiral after carbonization, ready to have a bulb blown over it.

I claim as my invention--
1. An electric lamp for giving light by incandescence, consisting of a filament of carbon of high resistance, made as described, and secured to metallic wires, as set forth.
2. The combination of carbon filaments with a receiver made entirely of glass and conductors passing through the glass, and from which receiver the air is exhausted, for the purposes set forth.
3. A carbon filament or strip coiled and connected to electric conductors so that only a portion of the surface of such carbon conductors shall be exposed for radiating light, as set forth.
4. The method herein described of securing the platina contact-wires to the carbon filament and carbonizing of the whole in a closed chamber, substantially as set forth.
Signed by me this 1st day of November, A. D. 1879.

THOMAS A. EDISON

Witnesses:
    S. L. Griffin,
    John F. Randolph

T. A. EDISON
Electric-Lamp

No. 223,898 — Patented Jan. 27, 1880.

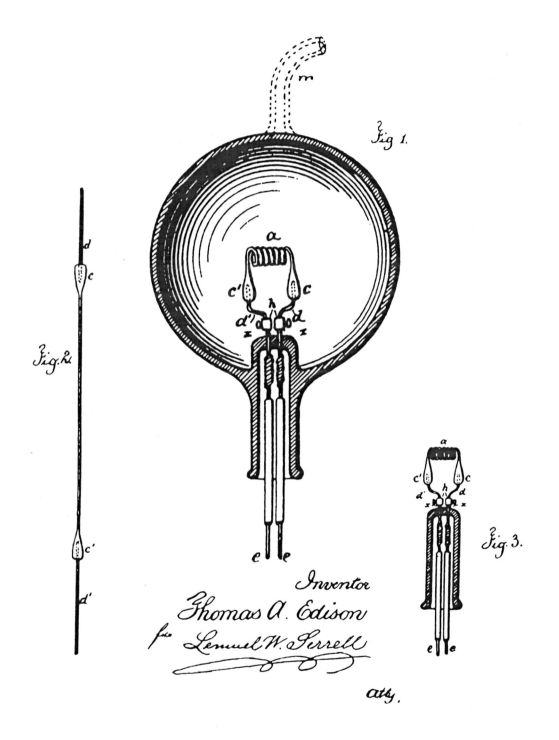

United States Patent Office

THOMAS A. EDISON, OF MENLO PARK, NEW JERSEY

IMPROVEMENT IN PHONOGRAPH OR SPEAKING MACHINES.

SPECIFICATION forming part of Letters Patent No. 200,521, dated February 19, 1878

Application filed December 24, 1877

To all whom it may concern:

Be it known that I, THOMAS A. EDISON, of Menlo Park, in the county of Middlesex and State of New Jersey, have invented an Improvement in Phonograph or Speaking Machines, of which the following is a specification:

The object of this invention is to record in permanent characters the human voice and other sounds, from which characters such sound may be reproduced and rendered audible again at a future time.

The invention consists in arranging a plate, diaphragm, or other flexible body capable of being vibrated by the human voice or other sounds, in conjunction with a material capable of registering the movements of such vibrating body by embossing or indenting or altering such material, in such a manner that such register-marks will be sufficient to cause a second vibrating plate or body to be set in motion by them, and thus reproduce the motions of the first vibrating body.

The invention further consists in the various combinations of mechanism to carry out my invention.

I have discovered, after a long series of experiments, that a diaphragm or other body capable of being set in motion by the human voice does not give, except in rare instances, superimposed vibrations, as has heretofore been supposed, but that each vibration is separate and distinct, and therefore it becomes possible to record and reproduce the sounds of the human voice.

In the drawings, Figure 1 is a vertical section, illustrating my invention, and Fig. 2 is a plan of the same.

A is a cylinder having a helical indenting-groove cut from end to end--say, ten grooves to the inch. Upon this is placed the material to be indented, preferably metallic foil. This drum or cylinder is secured to a shaft X, having at one end a thread cut with ten threads to the inch, the bearing P also having a thread cut in it.

L is a tube, provided with a longitudinal slot, and it is rotated by the clock-work at M, or other source of power.

The shaft X passes into the tube L, and it is rotated by a pin, 2, secured to the shaft, and passing through the slot on the tube L, the object of the long slot being to allow the shaft X to pass endwise through the center or support P by the action of the screw on X. At the same time that the cylinder is rotated it passes toward the support O.

B is the speaking-tube or mouth-piece, which may be of any desired character, so long as proper slots or holes are provided to re-enforce the hissing consonants. Devices to effect this object are shown in my application, No. 143, filed August 28, 1877. Hence they are not shown or further described herein.

Upon the end of the tube or mouth-piece is a diaphragm, having an indenting-point of hard material secured to its center, and so arranged in relation to the cylinder A that the point will be exactly opposite the groove in the cylinder at any position the cylinder may occupy in its forward rotary movement.

The speaking-tube is arranged upon a standard, which, in practice, I provide with devices for causing the tube to approach and recede from the cylinder.

The operation of recording is as follows: The cylinder is, by the action of the screw in X, placed adjacent to the pillar P, which brings the indenting-point of the diaphragm G opposite the first groove on the cylinder, over which is placed a sheet of thick metallic foil, paper, or other yielding material. The tube B is then adjusted toward the cylinder until the indenting-point touches the material and indents it slightly. The clock-work is then set running, and words spoken in the tube B will cause the diaphragm to take up every vibration, and these movements will be recorded with surprising accuracy by indentations in the foil.

After the foil on the cylinder has received the required indentations, or passed to its full limit toward O, it is made to return to P by proper means, and the indented material is brought to a position for reproducing and rendering audible the sound that had been made by the person speaking into the tube B.

C is a tube similar to B, except that the diaphragm is somewhat lighter and more sensitive, although this is not actually necessary. In front of this diaphragm is a light spring, D, having a small point shorter and finer than the indenting-point on the diaphragm of B. This spring and point are so arranged as to fall exactly into the path of all the indentations. This spring is connected to the diaphragm F of C by a thread or other substance capable of conveying the movements of D. Now, when the cylinder is allowed to rotate, the spring D is set in motion by each indentation corresponding to its depth and length. This motion is conveyed to the diaphragm either by vibrations through a thread or directly by connecting the spring to the diaphragm F, and these motions being due to the indentations, which are an exact record of every movement of the first diaphragm, the voice of the speaker is reproduced exactly and clearly, and with sufficient volume to be heard at some distance.

The indented material may be detached from the machine and preserved for any length of time, and by replacing the foil in a proper manner the original speaker's voice can be reproduced, and the same may be repeated frequently, as the foil is not changed in shape if the apparatus is properly adjusted.

The record, if it be upon tin-foil, may be stereotyped by means of the plaster-of-paris process, and from the stereotype multiple copies may be made expeditiously and cheaply by casting or by pressing tin foil or other material upon it. This is valuable when musical compositions are required for numerous machines.

It is obvious that many forms of mechanism may be used to give motion to the material to be indented. For instance, a revolving plate may have a volute spiral cut both on its upper and lower surfaces, on the top of which the foil or indenting material is laid and secured in a proper manner. A two-part arm is used with this disk, the portion beneath the disk having a point in the lower groove, and the portion above the disk carrying the speaking and receiving diaphragmic devices, which arm is caused, by the volute spiral groove upon the lower surface, to swing gradually from near the center to the outer circumference of the plate as it is revolved, or vice versa.

An apparatus of this general character adapted to a magnet that indents the paper is shown in my application for a patent, No. 128, filed March 26, 1877; hence no claim is made herein to such apparatus, and further description of the same is unnecessary.

A wide continuous roll of material may be used, the diaphragmic devices being reciprocated by proper mechanical devices backward and forward over the roll as it passes forward; or a narrow strip like that in a Morse register may be moved in contact with the indenting-point, and from this the sounds may be reproduced. The material employed for this purpose may be soft paper saturated or coated with paraffine or similar material, with a sheet of metal foil on the surface thereof to receive the impression from the indenting-point.

I do not wish to confine myself to reproducing sound by indentations only, as the transmitting or recording device may be in a sinuous form, resulting from the use of a thread passing with paper beneath the pressure-rollers *t*, (see Fig. 3), such thread being moved laterally by a fork or eye adjacent to the roller *t*, and receiving its motion from the diaphragm G, with which such fork or eye is connected, and thus record the movement of the diaphragm by the impression of the thread in the paper to the right and left of a straight line, from which indentation the receiving-diaphragm may receive its motion and the sound be reproduced, substantially in the manner I have already shown; or the diaphragm may, by its motion, give more or less pressure to an inking-pen, *u*, Fig. 4, the point of which rests upon paper or other material moved along regularly beneath the point of the pen, thus causing more or less ink to be deposited upon the material, according to the greater or lesser movement of the diaphragm. These ink-marks serve to give motion to a second diaphragm when the paper containing such marks is drawn along beneath the end of a lever resting upon them and connected to such diaphragm, the lever and diaphragm being moved by the friction between the point being greatest, or the thickness of the ink being greater where there is a large quantity of ink than where there is a small quantity. Thus the original sound-vibrations are reproduced upon the second diaphragm.

I claim as my invention--

1. The method herein specified of reproducing the human voice or other sounds by causing the sound-vibrations to be recorded, substantially as specified, and obtaining motion from that record, substantially as set forth, for the reproduction of the sound-vibrations.

2. The combination, with a diaphragm exposed to sound-vibrations, of a moving surface of yielding material--such as metallic foil--upon which marks are made corresponding to the sound-vibrations, and of a character adapted to use in the reproduction of the sound, substantially as set forth.

3. The combination, with a surface having marks thereon corresponding to sound-vibrations, of a point receiving motion from such marks, and a diaphragm connected to said point, and responding to the motion of the point, substantially as set forth.

4. In an instrument for making a record of sound-vibrations, the combination, with the diaphragm and point, of a cylinder having a helical groove and means for revolving the cylinder and communicating an end movement corresponding to the inclination of the helical groove, substantially as set forth.

Signed by me this 15th day of December, A. D. 1877.

THOS. A. EDISON

Witnesses:
    Geo. T. Pinckney,
    Chas. H. Smith

T. A. EDISON
Phonograph or Speaking Machine

No. 200,521 — Patented Feb. 19, 1878.

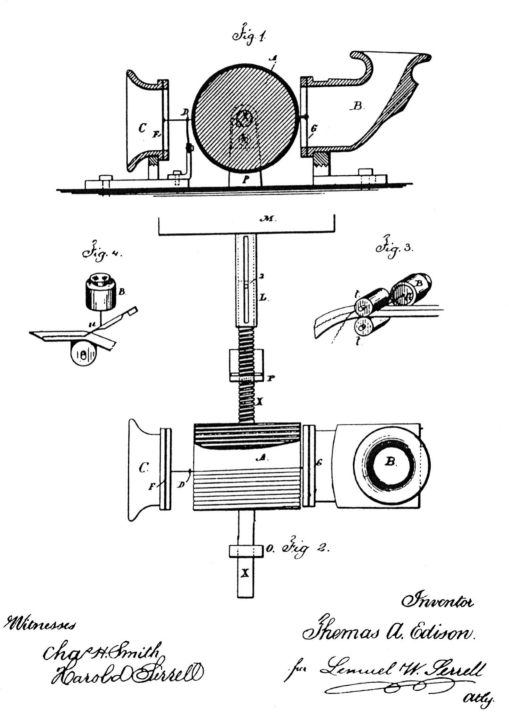

BENJAMIN FRANKLIN was born in Boston, Massachusetts Bay Colony in 1706 and died in Philadelphia, Pennsylvania in 1790. Famous as a statesman, publisher, author, printer, inventor, and scientist. Electricity was one of his many interests. He is credited with inventing the lightning rod.

Following are Franklin's two most notable electrical articles. These can be found reprinted in the book by Lemay, J. A. (editor), Franklin, (New York: Literary Classics of the United States, Inc., 1987), p.374-75 & 600-04.

"THE KITE EXPERIMENT"

This article was written by Franklin for The Pennsylvania Gazette, October 19, 1752.

<Don't try to do this very dangerous kite experiment, modern electrical engineers are in agreement that it was a miracle that Franklin survived.-- S. Tubbs, Editor>

As frequent Mention is made in the News Papers from *Europe*, of the Success of the *Philadelphia* Experiment for drawing the Electric Fire from Clouds by Means of pointed Rods of Iron erected on high Buildings, etc. it may be agreeable to the Curious to be inform'd, that the same Experiment has succeeded in *Philadelphia*, tho' made in a different and more easy Manner, which any one may try, as follows.

Make a small Cross of two light strips of Cedar, the Arms so long as to reach to the four Corners of a large thin Silk Handkerchief when extended; tie the Corners of the Handkerchief to the Extremities of the Cross, so you have the Body of a Kite; which being properly accommodated with a Tail, Loop and String, will rise in the Air, like those made of Paper; but this being of Silk is fitter to bear the Wet and Wind of a Thunder Gust without tearing. To the Top of the upright Stick of the Cross is to be fixed a very sharp pointed Wire, rising a Foot or more above the Wood. To the End of the Twine, next to the Hand, is to be tied a silk Ribbon, and where the Twine and the silk join, a Key may be fastened. This Kite is to be raised when a Thunder Gust appears to be coming on, and the Person who holds the String must stand within a Door, or Window, or under some Cover, so that the Silk Ribbon may not be wet; and Care must be taken that the Silk Ribbon may not touch the Frame of the Door or Window. As soon as any of the Thunder Clouds come over the Kite, the pointed Wire will draw the Electric Fire from them, and the Kite, with all the Twine will be electrified, and the loose Filaments of the Twine will stand out every Way, and be attracted by an approaching Finger.

And when the Rain has wet the Kite and Twine, so that it can conduct the Electric Fire freely, you will find it stream out plentifully from the Key on the Approach of your Knuckle. At this Key the Phial may be charg'd; and from Electric Fire thus obtain'd, Spirits may be kindled, and all the other Electric Experiments be perform'd, which are usually done by the Help of a rubbed Glass Globe or Tube; and thereby the *Sameness* of the Electric Matter with that of Lightning compleatly demonstrated.

"OF LIGHTNING, AND THE METHOD (NOW USED IN AMERICA) OF SECURING BUILDINGS AND PERSONS FROM ITS MISCHIEVOUS EFFECTS"

This article was written by Franklin for <u>Paris</u>, Sept. 1767

Experiments made in electricity first gave philosophers a suspicion that the matter of lightning was the same with the electric matter. Experiments afterwards made on lightning obtained from the clouds by pointed rods, received into bottles, and subjected to every trial, have since proved this suspicion to be perfectly well founded; and that whatever properties we find in electricity, are also the properties of lightning.

This matter of lightning, or of electricity, is an extream subtile fluid, penetrating other bodies, and subsisting in them, equally diffused.

When by an operation of art or nature, there happens to be a greater proportion of this fluid in one body than in another, the body which has the most, will communicate to that which has least, till the proportion becomes equal; provided the distance between them be not too great; or, if it is too great, till there be proper conductors to convey it from one to the other.

If the communication be through the air without any conductor, a bright light is seen between the bodies, and a sound is heard. In our small experiments we call this light and sound the electric spark and snap; but in the great operations of nature, the light is what we call *lightning*, and the sound (produced at the same time, tho' generally arriving later at our ears that the light does to our eyes) is, with its echoes, called *thunder*.

If the communication of this fluid is by a conductor, it may be without either light or sound, the subtle fluid passing in the substance of the conductor.

If the conductor be good and of sufficient bigness, the fluid passes through it without hurting it. If otherwise, it is damaged or destroyed.

All metals and water, are good conductors.--Other bodies may become conductors by having some quantity of water in them, as wood, and other materials used in building, but not having much water in them, they are not good conductors, and therefore are often damaged in the operation.

Glass, wax, silk, wool, hair, feathers, and even wood, perfectly dry are non-conductors: that is, they resist instead of facilitating the passage of this subtle fluid.

When this fluid has an opportunity of passing through two conductors, one good, and sufficient, as of metal, the other not so good, it passes in the best, and will follow it in any direction.

The distance at which a body charged with this fluid will discharge itself suddenly, striking through the air into another body that is not charged, or not so highly charg'd, is different according to the quantity of the fluid, the dimensions and form of the bodies themselves, and the state of the air between them.--This distance, whatever it happens to be between any two bodies, is called their *striking distance*, as till they come within that distance of each other, no stroke will be made.

The clouds have often more of this fluid in proportion than the earth; in which case as soon as they come near enough (that is, with the striking distance) or meet with a conductor, the fluid quits them and strikes into the earth. A cloud fully charged with this fluid, if so high as to be beyond the striking distance from the earth, passes quietly with making noise or giving light; unless it meets with other clouds that have less.

Tall trees, and lofty buildings, as the towers and spires of churches, become sometimes conductors between the clouds and the earth; but not being good ones, that is, not conveying the fluid freely, they are often damaged.

Buildings that have their roofs covered with lead, or other metal, and spouts of metal continued from the roof into the ground to carry off the water, are never hurt by lightning, as whenever it falls on such a building, it passes in the metals and not in the walls.

When other buildings happen to be within the striking distance from such clouds, the fluid passes in the walls whether of wood, brick or stone, quitting the walls only when it can find better conductors near them, as metal rods, bolts, and hinges of windows or doors, gilding on wainscot, or frames of pictures; the silvering on the backs of looking-glasses; the wires for bells; and the bodies of animals, as containing watry fluids. And in passing thro' the house it follows the direction of these conductors, taking as many in it's way as can assist it in its passage, whether in a strait or crooked line, leaping from one to the other, if not far distant from each other, only rending the wall in the spaces where these partial good conductors are too distant from each other.

An iron rod being placed on the outside of a building, from the highest part continued down into the moist earth, in any direction strait or crooked, following the form of the roof or other parts of the building, will receive the lightning at its upper end, attracting it so as to prevent its striking any other part; and, affording it a good conveyance into the earth, will prevent its damaging any part of the building.

A small quantity of metal is found able to conduct a great quantity of this fluid. A wire no bigger than a goose quill, has been known to conduct (with safety to the building as far as the wire was continued) a quantity of lightning that did prodigious damage both above and below it; and probably larger rods are not necessary, tho' it is common in America, to make them of half and inch, some of three quarters, or an inch diameter.

The rod may be fastened to the wall, chimney, etc. with staples of iron.--The lightning will not leave the rod (a good conductor) to pass into the wall (a bad conductor), through those staples.--It would rather, if any were in the wall, pass out of it into the rod to get more readily by that conductor into the earth.

If the building be very large and extensive, two or more rods may be placed at different parts, for greater security.

Small ragged parts of clouds suspended in the air between the great body of clouds and the earth (like leaf gold in electrical experiments), often serve as partial conductors for the lightning, which proceeds from one of them to another, and by their help comes within the striking distance to the earth or a building. It therefore strikes through those conductors a building that would otherwise be out of the striking distance.

Long sharp points communicating with the earth, and presented to such parts of clouds, drawing silently from them the fluid they are charged with, they are then attracted to the cloud, and may leave the distance so great as to be beyond the reach of striking.

It is therefore that we elevate the upper end of the rod six or eight feet above the highest part of the building, tapering it gradually to a fine sharp point, which is gilt to prevent its rusting.

Thus the pointed rod either prevents a stroke from the cloud, or, if a stroke is made, conducts it to the earth with safety to the building.

The lower end of the rod should enter the earth so deep as to come at the moist part, perhaps two or three feet; and if bent when under the surface so as to go in a horizontal line six or eight feet from the wall, and then bent again downwards three or four feet, it will prevent damage to any of the stones of the foundation.

A person apprehensive of danger coming from lightning, happening during the time of thunder to be in a house not so secured, will do well to avoid sitting near the chimney, near a looking glass, or any gilt pictures or wainscot; the safest place is in the middle of the room, (so it be not under a metal lustre suspended by a chain) sitting on one chair and laying the feet up in another. It is still safer to bring two or three mattrasses or beds into the middle of the room, and folding them up double, place the chair upon them; for they not being so good conductors as the wall, the lightning will not chuse an interrupted course through the air of the room and the bedding, when it can go thro' a continued better conductor the wall. But where it can be had, a hamock or swinging bed, suspended by silk cords equally distant from the walls on every side, and from the ceiling and floor above and below, affords the safest situation a person can have in any room whatever; and what indeed may be deemed quite free from danger of any stroke by lightning.

JOSEPH HENRY was born in Albany, New York in 1797 and died in Washington, D.C. in 1878. After Franklin, he was one of America's first great scientists. He is best known for his discovery of self-induction in 1830. Henry discovered electromagnetic induction in 1830, a year before Faraday published and received credit for the discovery. Unfortunately, Henry did not publish his findings before Faraday. Henry was the first secretary of the Smithsonian Institution in Washington, D.C. and one of Lincoln's chief technical advisers during the U.S. civil war.

Following is an article Joseph Henry wrote for Silliman's American Journal of Science, July 1832, Vol. XXII, p. 403-08; Scientific Writings, Vol. I, p. 73. A copy of it appears in the book by Lawn, R. (editor), Electro-magnetism, (New York: Arno Press, 1981), p.3-9.

"ON THE PRODUCTION OF CURRENTS AND SPARKS OF ELECTRICITY FROM MAGNETISM"

ALTHOUGH the discoveries of Oersted, Arago, Faraday, and others, have placed the intimate connection of electricity and magnetism in a most striking point of view, and although the theory of Ampère has referred all the phenomena of both these departments of science to the same general laws, yet until lately one thing remained to be proved by experiment, in order more fully to establish their identity; namely the possibility of producing electrical effects from magnetism. It is well known that surprising magnetic results can readily be obtained from electricity, and at first sight it might be supposed that electrical effects could with equal facility be produced from magnetism; but such has not been found to be the case, for although the experiment has often been attempted, it has nearly as often failed.

It early occurred to me that if galvanic magnets on my plan were substituted for ordinary magnets, in researches of this kind, more success might be expected. Besides their great powers these magnets possess other properties, which render them important instruments in the hands of the experimenter; their polarity can be instantaneously reversed, and their magnetism suddenly destroyed or called into full action, according as the occasion may require. With this view, I commenced, last August, the construction of a much larger galvanic magnet than, to my knowledge, had before been attempted, and also made preparations for a series of experiments with it on a large scale, in reference to the production of electricity from magnetism. I was, however, at that time accidentally interrupted in the prosecution of these experiments, and have not been able since to resume them, until within the last few weeks, and then on a much smaller scale than was first intended. In the meantime, it has been announced in the 117th number of the Library of Useful Knowledge, that the result so much sought after has at length been found by Mr. Faraday of the Royal Institution. It states that he has established the general fact, that when a piece of metal is moved in any direction, in front of a magnetic pole, electrical currents are developed in the metal, which pass in a direction at right angles to its own motion, and also that the application of this principle affords a complete and satisfactory explanation of the phenomena of magnetic rotation. No detail is given of the experiments, and it is somewhat surprising that results so interesting, and which certainly form a new era in the history of electricity and magnetism, should not have been more fully described before this time in some of the English publications; the only mention I have found of them is the following short account from the Annals of Philosophy for April, under the head of Proceedings of the Royal Institution:

"Feb. 17.--Mr. Faraday gave an account of the first two parts of his researches in electricity; namely, Volta-electric induction and magneto-electric induction. If two wires, A and B, be placed side by side, but not in contact, and a Voltaic current be passed through A, there is instantly a current produced by induction in B, in the opposite direction. Although the principal current in A be continued, still the secondary current in B is not found to accompany it, for it ceases after the first moment, but when the principal current is stopped then there is a second current produced in B, in the opposite direction to that of the first produced by the inductive action, or in the same direction as that of the principal current."

"If a wire, connected at both extremities with a galvanometer, be coiled in the form of a helix around a magnet, no current of electricity takes place in it. This is an experiment which has been made by various persons hundreds of times, in the hope of evolving electricity from magnetism, and as in other cases in which the wishes of the experimenter and the facts are opposed to each other, has given rise to very conflicting conclusions. But if the magnet be withdrawn from or introduced into such a helix, a current of electricity is produced *whilst the magnet is in motion*, and is rendered evident by the deflection of the galvanometer. If a single wire be passed by a magnetic pole, a current* of electricity is induced through it which can be rendered sensible."

Before having any knowledge of the method given in the above account, I had succeeded in producing electrical effects in the following manner, which differs from that employed by Mr. Faraday, and which appears to me to develop some new and interesting facts. A piece of copper wire, about thirty feet long and covered with elastic varnish, was closely coiled around the middle of the soft iron armature of the galvanic magnet described in Vol. XIX of the American Journal of Science, and which, when excited, will readily sustain between six hundred and seven hundred pounds. The wire was wound upon itself as to occupy

---

* Phil. Mag., and Annals of Philosophy, April, 1832; vol. XI, p. 300.

only about one inch of the length of the armature which is seven inches in all. The armature, thus furnished with wire, was placed in its proper position across the ends of the galvanic magnet, and there fastened so that no motion could take place. The two projecting ends of the helix were dipped into two cups of mercury, and there connected with a distant galvanometer by means of two copper wires, each about forty feet long. This arrangement being completed, I stationed myself near the galvanometer and directed an assistant at a given word to immerse suddenly, in a vessel of dilute acid, the galvanic battery attached to the magnet. At the instant of immersion, the north end of the needle was deflected $30°$ to the west, indicating a current of electricity from the helix surrounding the armature. The effect however appeared only as a single impulse, for the needle, after a few oscillations, resumed its former undisturbed position in the magnetic meridian, although the galvanic action of the battery, and consequently the magnetic power was still continued. I was, however, much surprised to see the needle suddenly deflected from a state of rest to about $20°$ to the east, or in a contrary direction when the battery was withdrawn from the acid, and again deflected to the west when it was re-immersed. This operation was repeated many times in succession, and uniformly with the same result, the armature the whole time remaining immovably attached to the poles of the magnet, no motion being required to produce the effect, as it appeared to take place only in consequence of the instantaneous development of the magnetic action in one, and the sudden cessation of it in the other.*

This experiment illustrates most strikingly the reciprocal action of the two principles of electricity and magnetism, if indeed it does not establish their absolute identity. In the first place, magnetism is developed in the soft iron of the galvanic magnet by the action of the currents of electricity from the battery, and secondly the armature, rendered magnetic by contact with the poles of the magnet, induces in its turn currents of electricity in the helix which surrounds it; we have thus as it were electricity converted into magnetism and this magnetism again into electricity.

Another fact was observed which is somewhat interesting inasmuch as it serves in some respects to generalize the phenomena. After the battery had been withdrawn from the acid, and the needle of the galvanometer suffered to come to a state of rest after the resulting deflection, it was again deflected in the same direction by partially detaching the armature from the poles of the magnet to which it continued to adhere from the action of the residual magnetism, and in this way, a series of deflections, all in the same direction, was produced by merely slipping off the armature by degrees until the contact was entirely broken. The following extract from the register of the experiments exhibits the relative deflections observed in one experiment of this kind.

* [*This experiment was performed, at the latest, in August 1831, and most probably in August 1830*]

At the instant of immersion of the battery, deflection	40° west.
At the instant of emersion of the battery,	"	18° east.
Armature partially detached,	"	7° east.
Armature entirely detached,	"	12° east.

The effect was reversed in another experiment, in which the needle was turned to the west in a series of deflections by dipping the battery but a small distance into the acid at first and afterwards immersing it by degrees.

From the foregoing facts it appears that a current of electricity is produced, for an instant, in a helix of copper wire surrounding a piece of soft iron whenever magnetism is induced in the iron; and a current in an opposite direction when the magnetic action ceases; also that an instantaneous current in one or the other direction accompanies every change in the magnetic intensity of the iron.

Since reading the account before given of Mr. Faraday's method of producing electrical currents I have attempted to combine the effects of motion and induction; for this purpose a rod of soft iron ten inches long and one inch and a quarter in diameter, was attached to a common turning lathe, and surrounded with four helices of copper wire in such a manner that could be suddenly and powerfully magnetized, while in rapid motion, by transmitting galvanic currents through three of the helices; the fourth being connected with the distant galvanometer was intended to transmit the current of induced electricity; all the helices were stationary while the iron rod revolved on its axis within them. From a number of trials in succession, first with the rod in one direction, then in the opposite, and next in a state of rest, it was concluded that no perceptible effect was produced on the intensity of the *magneto-electric* current by a rotary motion of the iron combined with its sudden magnetization.

The same apparatus, however, furnished the means of measuring separately the relative power of motion and induction in producing electrical currents. The iron rod was first magnetized by currents through the helices attached to the battery and while in this state one of its ends was quickly introduced into the helix connected with the galvanometer; the deflection of the needle in this case was seven degrees. The end of the rod was next introduced into the same helix while in its natural state and then suddenly magnetized; the deflection in this instance amounted to thirty degrees, showing a great superiority in the method of induction.

The next attempt was to increase the *magneto-electric* effect while the magnetic power remained the same, and in this I was more successful. Two iron rods, six inches long and one inch in diameter, were each surrounded by two helices and then placed perpendicularly on the face of the armature, and between it and the poles of the magnet, so that each rod formed as it were a prolongation of the poles, and to these the armature adhered when the magnet was excited. With this arrangement, a current from one helix produced a deflection of thirty-seven degrees; from two helices both on the same rod fifty-two degrees, and from three fifty-nine degrees; but when four helices were used the deflection was only fifty-five degrees, and when to these were added the helix of smaller wire around the armature, the deflection was no more than thirty degrees. This result may perhaps have been somewhat affected by the want of proper insulation in the several spires of the helices; it however establishes the fact that an increase in the electric current is produced by using at least two or three helices instead of one. The same principle was applied to another arrangement which seems to afford the maximum of electric development from a given magnetic power; in place of the two pieces of iron and the armature used in the last experiments, the poles of the magnet were connected by a single rod of iron, bent into the form of a horse-shoe, and its extremities filed perfectly flat so as to come in perfect contact with the faces of the poles; around the middle of the arch of this horse-shoe, two strands of copper wire were tightly coiled one over the other. A current from one of these helices deflected the needle one hundred degrees, and when both were used the needle was deflected with such force as to make a complete circuit. But the most surprising effect was produced when instead of passing the current, through the long wires to the galvanometer, the opposite ends of the helices were held nearly in contact with each other, and the magnet suddenly excited; in this case a small but vivid spark was seen to pass between the ends of the wires, and this effect was repeated as often as the state of intensity of the magnet was changed.

In these experiments the connection of the battery with the wires from the magnet was not formed by soldering, but by two cups of mercury which permitted the galvanic action on the magnet to be instantaneously suspended and the polarity to be changed and rechanged without removing the battery from the acid; a succession of vivid sparks was obtained by rapidly interrupting and forming the communication by means of one of these cups; but the greatest effect was produced when the magnetism was entirely destroyed and instantaneously reproduced by a change of polarity.

It appears from the May number of the Annals of Philosophy that I have been anticipated in this experiment of drawing sparks from the magnet by Mr. James D. Forbes of Edinburgh, who obtained a spark on the 30th of March; my experiments being made during the last two weeks of June. A simple notification of his result is given, without any account of the experiment, which is reserved for a communication to the Royal Society of Edinburgh; my result is therefore entirely independent of his and was undoubtedly obtained by a different process.

Electrical self-induction in a long helical wire

I have made several other experiments in relation to the same subject, but which more important duties will not permit me to verify in time for this paper. I may however mention one fact which I have not seen noticed in any work, and which appears to me to belong to the same class of phenomena as those before described; it is this: when a small battery is moderately excited by diluted acid, and its poles which should be terminated by cups of mercury, are connected by a copper wire not more than a foot in length, no spark is perceived when the connection is either formed or broken; but if a wire thirty or forty feet long be used instead of the short wire, though no spark will be perceptible when the connection is made, yet when it is broken by drawing one end of the wire from its cup of mercury, a vivid spark is produced. If the action of the battery be very intense, a spark will be given by the short wire; in this case it is only necessary to wait a few minutes until the action partially subsides, and until no more sparks are given from the short wire; if the long wire be now substituted a spark will again be obtained. The effect appears somewhat increased by coiling the wire into a helix; it seems also to depend in some measure on the length and thickness of the wire. I can account for these phenomena only by supposing the long wire to become charged with electricity, which by its re-action on itself projects a spark when the connection is broken.*

* [*This experiment was performed in August 1829.*]

CHARLES PROTEUS STEINMETZ

was born in Breslau, Prussia in 1865 and died in Schenectady, New York in 1923. His ideas on alternating current systems helped begin the electrical era in the U. S. Much of his work was done for the General Electric Company. Steinmetz theoretically described magnetic hysteresis losses, developed new simple and efficient ways of doing calculations on alternating current circuits, and developed an electrical transient theory.

Following is the Presidential address by Charles Proteus Steinmetz given at the 19th Annual Convention of the American Institute of Electrical Engineers, Great Barrington, Mass., June 21, 1902. A copy of it appears in the book by Alger, P. and Caldecott, E. (editors), <u>Steinmetz the Philosopher</u>, (Schenectady, New York: Mohawk Development Services, Inc., 1965), p.136-141.

"THE EDUCATION OF ELECTRICAL ENGINEERS"

Gentlemen: With today's meeting ends the first Institute year in the new century, and it may then be appropriate to look back into the work achieved and forward into the future.

At the entrance of the 19th century we see empirical science struggling for recognition against metaphysical speculation. At the entrance of the 20th century metaphysics has practically ceased to be considered, and empirical science is universally acknowledged as the source of all human progress. In gradual advance from success to success this victory has been won. Empirical science can now expand without any restraint. But with the removal of the opposition of a hostile philosophy, has been removed also the searching criticism into the correctness of the results, the methods and theories propounded by empirical science, and thereby the absolute safety which characterized the work of the earlier period. Herein lies the greatest danger to the unbroken progress of science.

Facts the Foundation

Facts, whether observed incidentally or intentionally by experiment, are the foundation of science. But the mass of facts has accumulated so as to be beyond comprehension, and we are forced to work with the conclusions from the facts, the empirical rule, theory and hypothesis, and have become so used thereto that gradually the dividing line between the theory and the underlying facts has become blurred, so that we attribute to the theory the same certainty as the facts on which it is based. But, however well a theory represents the fact and proves its value in successful application, it unavoidably contains besides the facts on which it is founded, foreign elements, mainly the personal equation of its originator and the personal equation of its times. When, building upon theories, the danger exists that instead of building upon the facts represented by the theory, we may build upon the personal equation of the time, without realizing the unsafe foundation of our structure.

The reverence for facts as the foundation of science has been extended to the theories based upon them. But when, by building theories upon theories, conclusions are derived which cease to be intelligible, it appears time to search into the foundations of the structure and to investigate how far the facts really warrant the conclusions. If we are told of matter moving in a vacuum tube with velocity comparable with that of light, of small chips breaking off atoms, of the free atoms of chlorine and sodium floating around in a salt solution charged with opposite electric charges, whatever that may be, of electricity being propagated not through the conductor, as we used to assume, but through the space surrounding it, etc., it is time to pause and try to understand.

In the gradual replacement of facts by the belief in theories more or less inadequately representing the facts, lies the chief danger to further scientific progress. Here is a field where splendid work in destructive criticism can and should be done by the younger generation, which, after leaving college with all the theoretical armament required, is not yet handicapped by personal relations with the men whose names are identified with the ruling theories.

Especially for our branch of electrical engineering, this is of fundamental importance, since all the success of our work depends upon the absolute safety of the foundations on which we build.

The Younger Generation

All future progress in science and engineering depends upon the young generation, and to ensure an unbroken advance it is of preeminent importance that the coming generation enters the field properly fitted for the work. Here the outlook appears to me by no means entirely encouraging.

It is not the object of the college or university to turn out full-fledged engineers who can handle any engineering problem of any magnitude. The available time is altogether too short for this, and a very great deal of practical experience is required which is not available to the educational institution. It is not necessary, either, and the college graduate is not expected to take immediate charge of engineering works of great magnitude. All the educational institution can do and should do is to fit the student to take up the practical work as efficiently as possible, and to give him *a thorough understanding of the fundamental principles of electrical engineering and allied sciences*, and *a good knowledge of the methods* of dealing with engineering problems. At present the average college course does not do this.

One of the reasons for the inefficiency of the present college course is the competition between colleges. By each college trying to teach more than any other, the quantity of material taught has gradually increased so that it is no longer possible to give a thorough understanding, and memorizing takes the place of understanding. Memorizing, however, is an entirely useless waste of energy, since anything that is not perfectly understood, but merely memorized will be forgotten in a short time if not continuously applied, and if continuously applied it would be remembered anyway. If of the amount of material in electrical engineering as well as other branches which the educational institution of today attempts to teach, one-half or more should be dropped altogether, and the rest taught so as to be fully understood, with special reference to general principles and methods, the product of the institution would be far superior and more successful in practical life. To be dropped, then, are all formulas, rules, etc., beyond the most simple ones, of the scope of the multiplication table, etc. When needed, formulas can be derived from the fundamental principles or looked up in the literature, and their memorizing is a mere waste of valuable time, useless since the memory cannot retain what the understanding is unable to reproduce from its premises. This, for instance, applies to the various formulas of electro-magnetic induction, the equations of the transformer and other apparatus, and to theoretical investigations in general.

Still more than to electrical engineering this applies to the allied branches of science and engineering, of which the electrical engineer should have a fair understanding. Here anything beyond fundamental principles is objectionable. The electrical engineer should understand the principle of the steam engine, of the hydraulic turbine, should know why the spouting velocity and the peripheral speed of the waterwheel are related to each other theoretically, and how this relation is modified actually, but he should not memorize the mathematical theory of the turbine with its half-dozen unknown friction coefficients. He is not expected to design turbines, and if he did, his design would surely be a failure, but he is expected to understand the turbine, the steam engine, etc. The engineering graduate should know the principles of the strength of materials, the distribution of forces in the loaded beam, the reasons leading to the "I" beam, and to the triangle in bridge and roof construction, but time should not be wasted in calculating bridges or deducing elastic curves, etc.

In short, in electrical engineering, and still more in all allied branches, nothing beyond the general principles is needed for success, but the principles should be fully understood, and with the limited time at the disposition of the student, this cannot be done if time is wasted in memorizing things to be forgotten afterward.

Step-by-step Method

One objectionable feature of the instruction of most colleges is the step-by-step method. One subject is taken up, by application of sufficient time and energy pushed through, and then after passing an examination dropped to take up another subject. It is true that by any steady application to one subject a great deal can be learned and splendid results derived in the examination papers, but all that is learned in this manner is just as rapidly forgotten. To understand a matter thoroughly, so as really to have a lasting benefit from it, and not merely make a good showing in examination papers, requires several years' familiarity. Therefore, any subject that is not kept up during the whole college course might just as well be dropped altogether and the time spent therein saved.

We hear that college graduates have no mastery of language, cannot express themselves logically, and it is therefore recommended that more time be devoted to literature and instruction in the English language. But as long as we do not teach logic, but have the English language taught by philologists, more or less of the character of the florid newspaper style, that is to write very many words on a very small subject, no useful results can be expected. Logic should be taught by a scientist, and the principle of English teaching to engineers should be to enable them to explain a subject intelligently and logically, with as few words as possible. That is, Faraday's researches, and not the product of the successful novel writer should be the example aimed at.

The combination of text-book and home work offers a splendid chance for incompetent instructors to make an elegant showing in examination papers, and turn out a very inferior grade of men, who sometimes do not even know what real understanding means. Free lectures on subjects which the instructor himself thoroughly understands (which is by no means always the case now) make it possible to direct the course so that the student's understanding follows it, and in this case home work becomes superfluous, except in connection with the laboratory. Reviewing the lectures and filling out the gaps due to absence, etc., appears to me the only legitimate requirement on the student's time outside the lecture room.

The present method of examination, which consists in expecting the student to answer ten questions or so within a few hours, is faulty. It shows what the student has memorized, but not how far he understands it. Furthermore, the brilliant student who usually answers nine questions out of ten correctly, and would have answered the last if he had not made a slight mistake, displaced a decimal or so, in practical life may be utterly useless, since while doing very rapid work, his work is not reliable, while a slow man who, however, can absolutely rely on the correctness of his work, will be very successful in practice while hardly passing the college examination.

Ideal Engineering Course

What, then, should an ideal electrical engineering course contain?

Of mathematics, plane and some solid geometry, arithmetic, and a good knowledge of algebra. Plane trigonometry--no spherical trigonometry required--and a thorough understanding of analytical geometry and of calculus, but no memorizing of integral formulas, etc.; they can always be looked up in a book when needed.

A thorough knowledge of general physics, especially of the law of conservation of energy, which I am sorry to say is not yet an integral part of our thinking and of chemistry, especially theoretical chemistry and the chemical laboratory, in the latter freely using tables, etc.

The electrical laboratory work should be taken up right at the beginning. Even before taking up the theory of apparatus, the theoretical investigation of alternating currents, self-induction, transformers, etc., the student should have met these phenomena and handled the apparatus in the laboratory. Only after seeing the effects of self-induction, for instance, on the alternating current circuit in the laboratory, will the theory of self-induction have any meaning or make an impression on the student. What the student has seen practically in the laboratory he will then take up in the theoretical course, and understand it; he will learn to calculate and control, and afterward, going back once more to the laboratory, to apply it. It is here that the average college graduate is inferior to the practical man, and in spite of his handicap regarding theoretical knowledge, the latter frequently pushes ahead of the college graduate because of his superior understanding of the phenomena, based on his familiarity with them.

Design of electrical apparatus is of very secondary utility and rather objectionable, with the exception perhaps of some very simple apparatus. The considerations on which designs are based in the engineering departments of the manufacturing companies, and especially the very great extent to which judgement enters into the work of the designing engineer, make the successful teaching of designing impossible to the college. Far better is the reverse operation, the analytical investigation of existing apparatus, and more instructive to the student. That is, to derive from existing first class apparatus by measurement and test, the constants of the apparatus, and to calculate these constants and compare them with tests. Besides, a very small percentage of the college graduates enter the field of designing, but most of them will have to handle apparatus, and it is therefore very much more important to be familiar with the completed apparatus in all its characteristic features, in normal and in abnormal operation, than to be able to design it.

The preceding naturally can be a short outline only of the work which I consider appropriate to an electrical engineering course, but I believe the general principles of the course can be understood from it. They are to give the student all the ground work required to be successful in future practice, but not the impossible aim of a complete education as a practical engineer.

Following is an article Charles Proteus Steinmetz wrote for the <u>Harper's</u> magazine, January, 1922. A copy of it appears in the book by Alger, P. and Caldecott, E. (editors), <u>Steinmetz the Philosopher</u>, (Schenectady, New York: Mohawk Development Services, Inc., 1965), p.77-87.

"ELECTRICITY AND CIVILIZATION"

The chief characteristic of our age is the independence of man from his immediate surroundings. The savage necessarily must depend upon his immediate neighborhood for the necessities of life. Some barter and commerce developed during the barbarian ages, but in the absence of any efficient means of transportation, even up to fairly recent times, such commerce could deal with luxuries and rare articles only, but for the common necessities of life man was still dependent on his immediate surroundings, and a local crop failure meant famine and starvation.

The great French Revolution at the end of the 18th century made man politically free, changed him from a serf to a citizen, and so unfettered the ability, initiative and ambition of all. The invention of the steam engine advanced man from a machine doing the mechanical labor of the world to a machine tender, directing the machines capable of doing the work of thousands of men, and set him mechanically free. The higher intelligence and knowledge needed therefore required education of the masses of the people, and so gave them an intellectual freedom which the illiterate man of former ages could not have.

Thus came the great and rapid development of our modern industrial and engineering civilization, which is characterized by the almost complete independence of man from his surroundings. No matter where we live, whether in the center of the great metropolis or in a small village out in the wilderness, anything that man and earth produce anywhere is available to us. The mail takes the order at our house, and in due time, by steamship and railway train, by express company or mail, it is delivered at our house.

This development of the means of transportation of materials by the steamship lines which cover the oceans, the railways which cover the continents with a network of tracks, the system of express and mail service, has been the great achievement of the 19th century, the foundation of our civilization, as we are forcibly made to realize whenever the transportation system breaks down ever so little, as it did during the last years.

But civilization depends on two things: materials and energy. Equally needed with materials, from the necessities of life to its luxuries, is energy, or power as we often call it.

That is the thing which makes the wheels go around; which drives the factories and mills; which in the steam locomotive carries us far better and faster than our feet could; which in the rays of the electric light or the gas flame lights our homes and turns night into day; in the heat of coal burning in the stove warms our houses and makes our climate inhabitable; and in our homes fetches and carries, cools the air by the fan motor or cooks our food, drives the sewing machine or the ice cream freezer, sweeps and dusts by the vacuum cleaner, washes, irons, and does more and more of the manual labor, and can and will do still more in the future to make life agreeable and efficient.

But while the methods of supply, transportation and distribution of materials have been developed highly by the transportation system, which were the great work of the last century, in the energy supply for the needs of man we are still backward, and this is the present great limitation of our civilization which the engineer is endeavoring to overcome.

The Stores of Energy

We cannot make or create energy. Thus we have to take it from where we find it in nature and bring it where we need it. The two big stores of energy in nature are in the coal mine and the waterfall, the former supplying the chemical energy of fuel (coal, oil, natural gas, etc.), which is set free as heat energy by combustion, and the latter supplying the hydraulic or mechanical energy.

The first problem which we meet then is how to transport the energy from its source to the place where we need it. We can do this well enough with the chemical energy of coal, by carrying the coal in railway train or steamship. And so we are doing, though it is rather an inefficient way, as it costs more to bring the coal from the mine to the consumer than it does to mine it. But mechanical energy, as the hydraulic energy of water power, we cannot transport as much at all (or "transmit" as we usually call the transportation of energy), and before the advent of electrical energy transmission, the water powers have been practically useless. The only way was for the user of energy to locate at the water power. But the place where the water power is found rarely is suitable for an industry, hardly ever for a big city, and these are the two largest users of energy. It was the electrical engineer who made the water powers of use, by changing-"transforming"--the hydraulic energy of the waterfall into electrical energy, to send it over the electric transmission line to the distant places where energy is needed, and distribute it as electric energy.

There are only two kinds or forms in which energy can be economically carried over long distances--"transported" or "transmitted"--as the chemical energy of fuel by the railway car or steamship and as electrical energy by the transmission line. And when from your train window you see the coal cars going by or the electric transmission lines flying past, you realize that both fulfill the same function: carrying the energy, that is the power of doing things on which our civilization depends, from its source where it is found in nature to the place where we need it.

But while fuel energy and electrical energy both can be economically transported or transmitted, there is a vast difference in them when we arrive at the destination and meet the problem of distributing and transforming the energy into that form which we need: heat energy to warm our homes and cook our food; light energy to extend the hours of daylight; and mechanical energy to fetch and carry, to bring us from and to our work or pleasure, to turn the wheels of industry, to drive the motor--whether the small fan motor of a fraction of a cat's power which cools our room, or the giant motor in the steel mill which with the power of ten thousands of horses squashes steel ingots of tons of weight as if they were soft putty into the shape of rails to carry the train, or steel beams to support our building structures or to span the rivers as bridges.

We can change the chemical energy of fuel into heat, by burning it in our stoves and furnaces, in a fairly simple, though rather inefficient manner.

But when we wish to convert the fuel energy into mechanical power, we can do it efficiently and economically only in very large units, in the huge and highly complicated steam turbine stations of ten thousands or hundred thousands of horsepower, and we cannot mechanically distribute this energy except in a highly wasteful and inefficient manner, by shafts and belts and countershafts.

If we want light, we have to select special fuels, as kerosene, or first convert the fuel energy of coal into that of gas in gas works, and distribute the gas, and even then we are far from the convenience and cleanliness of the electric light.

Energy Distribution

It is the characteristic of electric energy that it can be distributed and converted into any other form of energy, in a very simple and highly efficient manner, and that the economy of distribution and conversion is practically the same, whether we want the minute amount of energy to ring our door bell, or with the power of hundred thousands of horses to drive the propellers of the battle cruiser.

I press the button, and the electric light flashes up. I close the switch, and the fan motor starts at my desk, or the elevator begins to move, carrying tons of load, or the giant electric locomotive starts pulling the thousand-ton train. And there is little difference in the efficiency of the small motor driving a sewing machine and the giant motor on the rolling mill of the steel plant; either gives, in mechanical power, practically the full amount of the electric power which it receives.

Electrical energy is unique in this respect, and it is the only form of energy which can be transmitted, distributed and converted into any other form or energy with high efficiency--that is, with losses which are almost negligible--in the simplest possible manner and with practically no attention. Closing the switch starts it, opening the switch again stops it, and that equally well and efficiently for the most minute power as for the largest amounts of power.

Electrical energy thus is the form of energy best suited for the transmission and supply of the world's energy demand--it is indeed the only form of energy capable of doing this. And when you see the electric transmission lines criss-crossing the country and spreading over it in a network of wires, just as during the last century the railways have spread their network of tracks over the country, you should realize that the electrical engineer is doing today for the world's energy supply what the railway engineer has done during the last century for the world's material supply; he is organizing the world's energy supply needed to complete and maintain our civilization.

Useful but Useless

Electric energy thus is the most useful form of energy--and at the same time it is the most useless. It is not found in nature in usable quantity; the electrical energy of the lightning flash is too erratic and too small in amount to make it worth while to collect it, even if it could be done, and all electric energy is produced by conversion from some other form of energy--mechanical in the generator, chemical in the battery. Electric energy is never used as such (except in minute amounts occasionally medically), but when used it is always first converted into some other form of energy. Thus electrical energy is the intermediary, when you want to take some form of energy from somewhere and deliver it as some other form of energy somewhere else. Electric energy is the only energy fitted for this function as intermediary, as carrier between the source and the user of energy, due to its ease, simplicity and efficiency of production from other forms of energy and conversion into other forms of energy, and the efficiency and economy of its transmission.

Thus, with the rare exception where a power user can locate at the waterfall, water powers are always converted into electrical energy and transmitted and distributed as such, and it was the development of electrical engineering which has opened up to the uses of man water power, the second largest source of energy.

The chemical energy of fuel, from coal mine, oil well or gas well, is still usually transmitted or transported as chemical energy, by railway train or steamship line. The proposition has been made and discussed to burn the coal at the mine under steam boilers, convert its energy into electrical energy, and transmit it as such. To some extent at least, this will undoubtedly be done in the future, as the major part of the mine coal is wasted by being so poor and mixed with dirt that it cannot be economically transported. It could, however, be burned in proper furnaces at the mine, and so made useful as electrical energy. The extent however to which we could hope to do this is limited by the limitation of the steam engine. The steam engine requires not only fuel to produce the steam, but also large amounts of water to condense the steam, and very often such condensing water is not available at the coal mine.

But while most of the fuel energy is still transmitted or transported as such, when it comes to the distribution of energy, more and more the electrical form of energy is used. That is, in a big electric station near the demand for power--the big city, mill or factory--the fuel energy is converted into electrical energy and distributed as such. We could distribute and deliver the energy as fuel, coal, oil or gas, but what then? There is no simple and efficient way to convert the energy of coal, etc., into mechanical energy to propel the trolley car, or drive the sewing machine, into light to light our homes, etc., such as afforded by the electric power.

The difference in the usefulness of electrical energy in deriving any other form of energy from it, compared with the energy of coal is best illustrated by such a simple convenience as the electric fan: push the button, and the fan starts; push it again, and it stops. Now imagine the problem of operating your desk fan by the energy of coal. You have attached to the fan a little steam engine, and to it a little boiler, and a little coal furnace, and when you want to start the fan, you start a fire in the little furnace on you desk, and get up steam in the little boiler, and operate the little steam engine to drive the fan. You see how impossible it is to use fuel energy for general energy distribution. You may say: "We would not use coal, but use gas or oil, in a gas engine. We would have a little gas engine attached to our fan." This is simpler, but you fill your room with the ill-smelling hot gases, and after all to keep the gas engine running you have to put a magneto for ignition, and this is larger than the whole electric fan motor. Or if you use battery ignition, the power you take out of the battery could by a small electric motor drive your fan.

Superiority of Electricity

This illustrates the superiority of electric power in energy distribution, and so a whole new industry has grown up in the last 25 years: the industry of electric power generation and distribution. From the small electric lighting stations of the early days have grown up huge electric power stations, some of them approaching a million horsepower, and more and more supplying all the energy demand of the city or country, whether for lighting homes or streets, for driving the surface trolleys or the rapid transit systems and the terminals of the steam railroads, or supplying energy to factories and mills--in short, taking care of the energy supply and distribution.

In the field of rail transportation, the electric motor has superseded all other means except the steam locomotive on our trunkline railways. But every engineer who has looked into the situation knows that the steam locomotive is doomed by its frightful wastefulness, and electrification in inevitable. By electric operation of our railways, even if all the electric power were generated by steam and no water power used, we would save about two-thirds of the coal now consumed by the locomotives--that is, hundred of millions of tons--and at the same time without a single mile of additional track, increase the capacity of our railroads by a quarter or more, due to the quicker start, better control and higher speed of the electric train.

In factories and mills, the electric motor is replacing the steam engine and thereby changing our industrial system. So we have seen in the last 25 years the cotton industry shifting from the New England States to the South, due to the economic advantage afforded by the abundant water powers of the southern streams.

When, in changing to electric power, we replace the steam engine by the electric motor, without any further change, this rarely is the most economical way, as the electric motor can do many things which the engine cannot do, and this permits a rearrangement of power supply, resulting in a great increase of economy. The steam engine is economical only in large units, and requires constant care and skilled attendance. Thus it is not possible to place a small steam engine at every machine where power is wanted; but one big steam engine drives the factory, through numerous shafts and counter-shafts and masses of belts, in which quite commonly more than half of the power is wasted, even when the factory is running full, and the waste becomes still greater when the factory is operated only partly, or when only a few machines have to be run for overtime work.

The electric motor, however, whether large or small, requires practically no attendance. Thus a separate motor may be attached to every machine, whether a sewing machine requiring a twentieth of a horsepower or a mill motor of a thousand horsepower. Thus all the shafting and belting disappear, light is let into the factory, and the safety vastly increased, and the enormous losses of power in the transmission saved.

So also in transportation: the electric locomotive is more efficient than the steam locomotive, but more efficient still is the electric motor car, and while steam trains have become larger and larger, to use the largest and most efficient locomotives, wherever possible, in electric traction individual motor cars are used, giving a more frequent and thus better service.

Do It Electrically

It therefore has been said: "To do a thing well and efficiently, do it electrically," and there is a great deal of truth in this.

In our households, and in general in everyday life, electricity is playing a larger and larger part.

Electricity is the only commodity which during the last 25 years has steadily decreased in price due to the rapid advance of electrical engineering, while everything else has increased. And even in the last few years, when all other commodities doubled and tripled in cost, the price of electricity has hardly increased at all, so that domestic uses of electricity which once were a luxury now have become more economical than the old ways, besides being far more convenient, cleanly and sanitary.

Thus electricity is supplying household power and saving labor, eliminating the drudgery which formerly made household work so unattractive, with fan motor and vacuum cleaner, the motor on the sewing machine or the ice cream freezer, the washing machine and ironing machine, the doorbell, and the electric flatiron. With electric cooking, from special services as electric toaster, coffee percolator, etc., to the electric range replacing the coal--or gas-fed cooking stove, electricity has found its field. And it is reasonable to expect that all the domestic and industrial work of the city, all locomotion and transportation, will some time be done by electricity and that in a not very distant future; all fires and combustion will be altogether forbidden by law within the city limits, as dangerous and unsanitary. It is not reasonable to believe our civilized society will forever permit filling the air and the sky above our cities with soot from a thousand smoke-belching chimneys, and poisoning the air of the city streets by the ill-smelling exhaust gases of thousands of gasoline cars, when electricity can perform the duty in a safer and better manner.

A Motor for Each Machine

As we have seen, it is not the best economy, in industrial electrification, merely to take out the steam engine and put an electric motor in its place, but best economy requires a rearrangement, the elimination of the mechanical power distribution by shafts and belts, and putting an electric motor at every machine. So in domestic electrification it would be hopelessly uneconomical, even with the lowest prices of electricity merely to replace the grate of the coal stove or the burner of the gas stove by an electric heater, due to the enormous waste of heat inevitable in the coal stove and even gas stove. But electric heat can be employed so much more directly, and with so little loss, as to make electric cooking economical--and often even cheaper than cooking by coal or gas stove--though the heat from electric power must always remain more expensive than heat by the combustion of coal.

This is most marked in domestic heating. Suppose we take out the grate from our hot air, steam or hot water heating furnace in the cellar of our house, put in an electric heater and try to use the present heating plant. Even at the lowest imaginable rate of electric power, the cost would be such as to make it economically hopeless. We get a better economy by putting an electric heater in every room, but still the cost of heating would inevitably be much greater than our present method, so that it could be used only in special cases, in the business centers of big cities, where the space occupied by the heating system is very valuable, or to heat some individual room, as a sick chamber or a bedroom, without the need of starting the entire heating plant; that is, it would be used as auxiliary to the coal furnace of today.

It is not possible that the cost of electric heat compared with that of coal can ever decrease sufficiently to make electric heating of our present houses generally economical. When electric energy is produced from coal, even in our most efficient huge steam turbine stations, we get only 15% to 20% of the energy of the coal as electric energy, due to the inherent limitation of the steam engine. If then this electric energy is used for heating, the coal used in producing the electric energy costs five to six times as much as the coal which would produce the same amount of heat directly by combustion. The cost of the fuel is only a part, often less than half, of the cost of the electric energy, so that the actual cost of the heat produced from electricity in domestic service must be 10 to 20 times as great as that of the cost of the same amount of heat produced by the burning of coal, and the price thus still higher.

There is no possibility that cheap hydraulic power could ever reduce the price of electricity so radically as to make the cost of electric heat comparable with that of heat from coal. Some people still believe that electricity from water power costs nothing or very little because no fuel is consumed in producing it, but the cost of fuel is only a part of the cost of electric power. A large part of the cost of power is the interest on development and depreciation of the plant. Hydro-electric plants almost invariably cost several times much as steam plants, due to the much more expensive and extensive hydraulic development, the cost of transmission lines, etc. Therefore, what is saved in the hydraulic station in the cost of fuel is in general pretty nearly lost in the higher cost of the development, with the result that electricity from water power can differ little in cost from that from steam power. That is, some water powers can produce electricity cheaper than the average steam station, and some large steam stations cheaper than the average water power station, and in general hydroelectric power is a little cheaper and a little less reliable than steam electric power, but the difference is not sufficient to give the water power any radical economic advantage.

We also must realize that if all the possible water powers were used--that is, every drop of rain which falls in the United States collected and its power converted into electricity--and all this electric power used for heating, the total amount of heat produced would be only about one-third as much as that given by our present coal consumption.

The House Without a Chimney

Nevertheless, even with the present prices of electric energy for domestic purposes, electric house heating might be economically feasible. But it means an entire rearrangement of the heating and ventilating methods, even of the construction of our buildings it means the "house without a chimney."

The walls of the house would be insulated against losses of heat by conduction through them; double or triple glass would be used in the windows; all the cracks and openings through which cold air might enter would be made perfectly tight; double or triple entrance doors would be used, not to lose appreciable heat when opening them. In this way, it would be possible to reduce the present losses of heat to a small fraction.

There remains the question of ventilation: usually, we let fresh--and cold--air in, and the foul warm air out, and so lose all the heat contained in the latter and over and over again have to heat new volumes of cold air. With the expensive electric heat, this is not permissible--and is not necessary--but a regenerative system of ventilation is used. That is, the heat contained in the foul warm air, which leaves the house, is transferred to the fresh cold air entering it, and so heats it, by the outgoing air passing around the pipes which carry in the fresh air. An air-tight house, and a regenerative system of ventilation, would reduce the amount of heat required to keep our homes warm during the cold season to such a small fraction of the heat required with our present method of building construction that electric heating would become economical with the present price of electricity. Such a change obviously can come gradually only.

NIKOLA TESLA was born a Serbian in Smiljan, Croatia in 1856 and died in New York City in 1943. He discovered the rotating magnetic field, the basis of most alternating current machinery. He also invented the Tesla coil, a high frequency, high voltage source of alternating current electricity. Tesla was a dreamer, poet, and genius. Electrical inventors still are looking through his works for clues to new inventions.

Following is an article Nikola Tesla wrote for the Thirtieth Anniversary of the Electrical World and Engineer, March 5, 1904. It can be found reprinted in the book by Childress D. H. (ed.), The Fantastic Inventions of Nikola Tesla, (Kempton, Illinois: Adventures Unlimited Press, 1993), p.219-40.

"TRANSMISSION OF ELECTRICAL ENERGY WITHOUT WIRES"

It is impossible to resist your courteous request extended on an occasion of such moment in the life of your journal. Your letter has vivified the memory of our beginning friendship, of the first imperfect attempts and undeserved successes, of kindnesses and misunderstandings. It has brought painfully to my mind the greatness of early expectations, the quick flight of time, and alas! the smallness of realizations. The following lines which, but for your initiative, might not have been given to the world for a long time yet, are an offering in the friendly spirit of old, and my best wishes for your future success accompany them.

Towards the close of 1898 a systematic research, carried on for a number of years with the object of perfecting a method of transmission of electrical energy through the natural medium, led me to recognize three important necessities: First, to develop a transmitter of great power; second, to perfect means for individualizing and isolating the energy transmitted; and, third, to ascertain the laws of propagation of currents through the earth and the atmosphere. Various reasons, not the least of which was the help proffered by my friend Leonard E. Curtis and the Colorado Springs Electric Company, determined me to select for my experimental investigations the large plateau, two thousand meters above sea-level, in the vicinity of that delightful resort, which I reached late in May, 1899. I had not been there but a few days when I congratulated myself on the happy choice and I began the task, for which I had long trained myself, with a grateful sense and full of inspiring hope. The perfect purity of the air, the unequaled beauty of the sky, the imposing sight of a high mountain range, the quiet and restfulness of the place--all around contributed to make the conditions for scientific observations ideal. To this was added the exhilarating influence of a glorious climate and a singular sharpening of the senses. In those regions the organs undergo perceptible physical changes. The eyes assume an extraordinary limpidity, improving vision; the ears dry out and become more susceptible to sound. Objects can be clearly distinguished there at distances such that I prefer to have them told by someone else, and I have heard--this I can venture to vouch for--the claps of thunder seven and eight hundred kilometers away. I might have done better still, had it not been tedious to wait for the sounds to arrive, in definite intervals, as heralded precisely by an electrical indicating apparatus--nearly an hour before.

In the middle of June, while preparations for other work were going on, I arranged one of my receiving transformers with the view of determining in a novel manner, experimentally, the electric potential of the globe and studying its periodic and casual fluctuations. This formed part of a plan carefully mapped out in advance. A highly sensitive, self-restorative device, controlling a recording instrument, was included in the secondary circuit, while the primary was connected to the ground and an elevated terminal of adjustable capacity. The variations of potential gave rise to electric surgings in the primary; these generated secondary currents, which in turn affected the sensitive device and recorder in proportion to their intensity. The earth was found to be, literally, alive with electrical vibrations, and soon I was deeply absorbed in the interesting investigation. No better opportunities for such observations as I intended to make could be found anywhere. Colorado is a country famous for the natural displays of electric force. In that dry and rarefied atmosphere the sun's rays beat the objects with fierce intensity. I raised steam, to a dangerous pressure, in barrels filled with concentrated salt solution, and the tin-foil coatings of some of my elevated terminals shriveled up in the fiery blaze. An experimental high-tension transformer, carelessly exposed to the rays of the setting sun, had most of its insulating compound melted out and was rendered useless. Aided by the dryness and rarefaction of the air, the water evaporates as in a boiler, and static electricity is developed in abundance. Lightning discharges are, accordingly, very frequent and sometimes of inconceivable violence. On one occasion approximately twelve thousand discharges occurred in two hours, and all in a radius of certainly less than fifty kilometers from the laboratory. Many of them resembled gigantic trees of fire with the trunks up or down. I never saw fire balls, but as compensation for my disappointment I succeeded later in determining the mode of their formation and producing them artificially.

In the latter part of the same month I noticed several times that my instruments were affected stronger by discharges taking place at great distances than by those near by. This puzzled me very much. What was the cause? A number of observations proved that it could not be due to the differences in the intensity of the individual discharges, and I readily ascertained that the phenomenon was not the result of a varying relation between the periods of my receiving circuits and those of the terrestrial disturbances. One night, as I was walking home with an assistant, meditating over these experiences, I was suddenly staggered by a thought. Years ago, when I wrote a chapter of my lecture before the Franklin Institute and the National Electric Light Association, it had presented itself to me, but I dismissed it as absurd and impossible. I banished it again. Nevertheless, my instinct was aroused and somehow I felt that I was nearing a great revelation.

It was on the third of July--the date I shall never forget-when I obtained the first decisive experimental evidence of a truth of overwhelming importance for the advancement of humanity. A dense mass of strongly charged clouds gathered in the west and towards the evening a violent storm broke loose which, after spending much of its fury in the mountains, was driven away with great velocity over the plains. Heavy and long persisting arcs formed almost in regular time intervals. My observations were now greatly facilitated and rendered more accurate by the experiences already gained. I was able to handle my instruments quickly and I was prepared. The recording apparatus being properly adjusted, its indications became fainter and fainter with the increasing distance of the storm, until they ceased altogether. I was watching in eager expectation. Surely enough, in a little while the indications again began, grew stronger and stronger and, after passing through a maximum, gradually decreased and ceased once more. Many times, in regularly recurring intervals, the same actions were repeated until the storm which, as evident from simple computations, was moving with nearly constant speed, had retreated to a distance of about three hundred kilometers. Nor did these strange actions stop then, but continued to manifest themselves with undiminished force. Subsequently, similar observations were also made by my assistant, Mr. Fritz Lowenstein, and shortly afterward several admirable opportunities presented themselves which brought out, still more forcibly, and unmistakably, the true nature of the wonderful phenomenon. No doubt, whatever remained: I was observing stationary waves.

As the source of disturbances moved away the receiving circuit came successively upon their nodes and loops. Impossible as it seemed, this planet, despite its vast extent, behaved like a conductor of limited dimensions. The tremendous significance of this fact in the transmission of energy by my system had already become quite clear to me. Not only was it practicable to send telegraphic messages to any distance without wires, as I recognized long ago, but also to impress upon the entire globe the faint modulations of the human voice, far more still, to transmit power, in unlimited amounts, to any terrestrial distance and almost without loss.

With these stupendous possibilities in sight, and the experimental evidence before me that their realization was henceforth merely a question of expert knowledge, patience and skill, I attacked vigorously the development of my magnifying transmitter, now, however, not so much with the original intention of producing one of great power, as with the object of learning how to construct the best one. This is, essentially, a circuit of very high self-induction and small resistance which in its arrangement, mode of excitation and action, may be said to be the diametrical opposite of a transmitting circuit typical of telegraphy by Hertzian or electromagnetic radiations. It is difficult to form an adequate idea of the marvelous power of this unique appliance, by the aid of which the globe will be transformed. The electromagnetic radiations being reduced to an insignificant quantity, and proper conditions of resonance maintained, the circuit acts like an immense pendulum, storing indefinitely the energy of the primary exciting impulses and impressions upon the earth of the primary exciting impulses and impressions upon the earth and its conducting atmosphere uniform harmonic oscillations of intensities which, as actual tests have shown, may be pushed so far as to surpass those attained in the natural displays of static electricity.

Simultaneously with these endeavors, the means of individualization and isolation were gradually improved. Great importance was attached to this, for it was found that simple tuning was not sufficient to meet the vigorous practical requirements. The fundamental idea of employing a number of distinctive elements, co-operatively associated, for the purpose of isolating energy transmitted, I trace directly to my perusal of Spencer's clear and suggestive exposition of the human nerve mechanism. The influence of this principle on the transmission of intelligence, and electrical energy in general, cannot as yet be estimated, for the art is still in the embryonic stage; but many thousands of simultaneous telegraphic and telephonic messages, through one single conducting channel, natural or artificial, and without serious mutual interference, are certainly practicable, while millions are possible. On the other hand, any desired degree of individualization may be secured by the use of a great number of co-operative elements and arbitrary variation of their distinctive features and order of succession. For obvious reasons, the principle will also be valuable in the extension of the distance of transmission.

Progress though of necessity slow was steady and sure, for the objects aimed at were in a direction of my constant study and exercise. It is, therefore, not astonishing that before the end of 1899 I completed the task undertaken and reached the results which I have announced in my article in the Century Magazine of June, 1900, every word of which was carefully weighed.

Much has already been done towards making my system commercially available, in the transmission of energy in small amounts of specific purposes, as well as on an industrial scale. The results attained by me have made my scheme of intelligence transmission, for which the name of "World Telegraphy" has been suggested, easily realizable. It constitutes, I believe, in its principle of operation, means employed and capacities of application, a radical and fruitful departure from what has been done heretofore. I have no doubt that it will prove very efficient in enlightening the masses, particularly in still uncivilized countries and less accessible regions, and that it will add materially to general safety, comfort and convenience, and maintenance of peaceful relations. It involves the employment of a number of plants, all of which are capable of transmitting individualized signals to the uttermost confines of the earth. Each of them will be preferably located near some important center of civilization and the news it receives through any channel will be flashed to all points of the globe. A cheap and simple device, which might be carried in one's pocket, may then be set up somewhere on sea or land, and it will record the world's news or such special messages as may be intended for it. Thus the entire earth will be converted into a huge brain, as it were, capable of response in every one of its parts. Since a single plant of but one hundred horse-power can operate hundreds of millions of instruments, the system will have a virtually infinite working capacity, and it must needs immensely facilitate and cheapen the transmission of intelligence.

The first of these central plants would have been already completed had it not been for unforeseen delays which, fortunately, have nothing to do with its purely technical features. But this loss of time, while vexatious, may, after all, prove to be a blessing in disguise. The best design of which I know has been adopted, and the transmitter will emit a wave complex of total maximum activity of ten million horse-power, one per cent. of which is amply sufficient to "girdle the globe." This enormous rate of energy delivery, approximately twice that of the combined falls of Niagara, is obtainable only by the use of certain artifices, which I shall make known in due course.

For a large part of the work which I have done so far I am indebted to the noble generosity of Mr. J. Pierpont Morgan, which was all the more welcome and stimulating, as it was extended at a time when those, who have since promised most, were the greatest of doubters. I have also to thank my friend, Stanford White, for much unselfish and valuable assistance. This work is now far advanced, and though the results may be tardy, they are sure to come.

Meanwhile, the transmission of energy on an industrial scale is not being neglected. The Canadian Niagara Power Company have offered me a splendid inducement, and next to achieving success for the sake of the art, it will give me the greatest satisfaction to make their concession financially profitable to them. In this first power plant, which I have been designing for a long time, I propose to distribute ten thousand horse-power under a tension of one hundred million volts, which I am now able to produce and handle with safety.

This energy will be collected all over the globe preferably in small amounts, ranging from a fraction of one to a few horse-power. One of its chief uses will be the illumination of isolated homes. I takes very little power to light a dwelling with vacuum tubes operated by high-frequency currents and in each instance a terminal a little above the roof will be sufficient. Another valuable application will be the driving of clocks and other such apparatus. These clocks will be exceedingly simple, will require absolutely no attention and will indicate rigorously correct time. The idea of impressing upon the earth American time is fascinating and very likely to become popular. There are innumerable devices of all kinds which are either now employed or can be supplied, and by operating them in this manner I may be able to offer a great convenience to the whole world with a plant of no more than ten thousand horse-power. The introduction of this system will give opportunities for invention and manufacture such as have never presented themselves before.

Knowing the far-reaching importance of this first attempt and its effect upon future development, I shall proceed slowly and carefully. Experience has taught me not to assign a term to enterprises the consummation of which is not wholly dependent on my own abilities and exertions. But I am hopeful that these great realizations are not far off, and I know that when this first work is completed they will follow with mathematical certitude.

When the great truth accidentally revealed and experimentally confirmed is fully recognized, that this planet, with all its appalling immensity, is to electric currents virtually no more than a small metal ball and that by this fact many possibilities, each baffling imagination and of incalculable consequence, are rendered absolutely sure of accomplishment; when the first plant is inaugurated and it is shown that a telegraphic message, almost as secret and non-interferable as a thought, can be transmitted to any terrestrial distance, the sound of the human voice, with all its intonations and inflections, faithfully and instantly reproduced at any other point of the globe, the energy of a waterfall made available for supplying light, heat or motive power, anywhere--on sea, or land, or high in the air--humanity will be like an ant heap stirred up with a stick: See the excitement coming!

Following is an article Nikola Tesla wrote for the Electrical Experimenter, May 1919. It appears that the Electrical Experimenter editors added some Tesla equipment photographs, drawings, and information describing them. All of the article's material is printed below.

"MY INVENTIONS; THE DISCOVERY OF THE TESLA COIL AND TRANSFORMER"

For a while I gave myself up entirely to the intense enjoyment of picturing machines and devising new forms. It was a mental state of happiness about as complete as I have ever known in life. Ideas came in an uninterrupted stream and the only difficulty I had was to hold them fast. The pieces of apparatus I conceived were to me absolutely real and tangible in every detail, even to the minutest marks and signs of wear. I delighted in imagining the motors constantly running, for in this way they presented to the mind's eye a more fascinating sight. When natural inclination develops into a passionate desire, one advances towards his goal in seven-league boots. In less than two months I evolved virtually all the types of motors and modifications of the system which are now identified with my name. It was, perhaps, providential that the necessities of existence commanded a temporary halt to this consuming activity of the mind. I came to Budapest prompted by a premature report concerning the telephone enterprise and, as irony or fate willed it, I had to accept a position as draftsman in the Central Telegraph Office of the Hungarian Government at a salary which I deem it my privilege not to disclose! Fortunately, I soon won the interest of the Inspector-in-Chief and was thereafter employed on calculations, designs and estimates in connection with new installations, until the Telephone Exchange was started, when I took charge of the same. The knowledge and practical experience I gained in the course of this work was most valuable and the employment gave me ample opportunities for the exercise of my inventive faculties. I made several improvements in the Central Station apparatus and perfected a telephone repeater or amplifier which was never patented or publicly described but would be creditable to me even today. In recognition of my efficient assistance the organizer of the undertaking, Mr. Puskas, upon disposing of his business in Budapest, offered me a position in Paris which I gladly accepted.

I never can forget the deep impression that magic city produced on my mind. For several days after my arrival I roamed thru the streets in utter bewilderment of the new spectacle. The attractions were many and irresistible, but, alas, the income was spent as soon as received. When Mr. Puskas asked me how I was getting along in the new sphere, I described the situation accurately in the statement that "the last twenty-nine days of the month are the toughest!" I led a rather strenuous life in what would now be termed "Rooseveltian fashion." Every morning, regardless of weather, I would go from the Boulevard St. Marcel, where I resided, to a bathing house on the Seine, plunge into the water, loop the circuit twenty-seven times and then walk an hour to reach Ivry, where the Company's factory was located. There I would have a woodchopper's breakfast at half-past seven o'clock and then eagerly await the lunch hour, in the meanwhile cracking hard nuts for the Manager of the Works, Mr. Charles Batchellor, who was an intimate friend and assistant of Edison. Here I

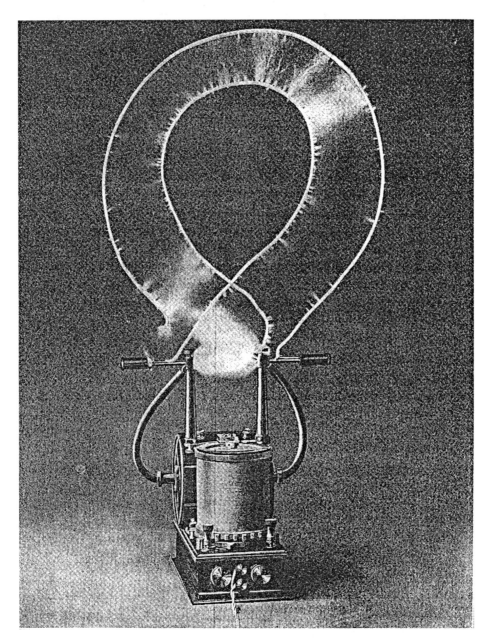

Fig.1-Tesla Oscillation Transformer (Tesla Coil) Presented by Lord Kelvin Before the British Association in August, 1897. This Small and Compact Instrument, Only 8 Inches High, Developed Two Square Feet of Streamers With Twenty Five Watts From the 110 Volt D.C. Supply Circuit. The Instrument Contains a Tesla Primary and Secondary, Condenser, and a Circuit Controller.

was thrown in contact with a few Americans who fairly fell in love with me because of my proficiency in--billiards. To these men I explained my invention and one of them, Mr. D. Cunningham, Foreman of the Mechanical Department, offered to form a stock company. The proposal seemed to me comical in the extreme. I did not have the faintest conception of what that meant except that it was an American way of doing things. Nothing came of it, however, and during the next few months I had to travel from one to another place in France and Germany to cure the ills of the power plants. On my return to Paris I submitted to one of the administrators of the Company, Mr. Rau, a plan for improving their dynamos and was given an opportunity. My success was complete and the delighted directors accorded me the privilege of developing automatic regulators which were much desired. Shortly after there was some trouble with the lighting plant which has been installed at the new railroad station in Strassburg, Alsace. The wiring was defective and on the occasion of the opening ceremonies a large part of a wall was blown out through a short-circuit right in the presence of old Emperor William I. The German Government refused to take the plant and the French Company was facing a serious loss. On account of my knowledge of the German language and past experience, I was entrusted with the difficult task of straightening out matters and early in 1883 I went to Strassburg on that mission.

The First Induction Motor Is Built.

Some of the incidents in that city have left an indelible record on my memory. By a curious coincidence, a number of men who subsequently achieved fame, lived there about that time. In later life I used to say, "There were bacteria of greatness in that old town. Others caught the disease but I escaped!" The practical work, correspondence, and conferences with officials kept me preoccupied day and night, but as soon as I was able to manage I undertook the construction of a simple motor in a mechanical shop opposite the railroad station, having brought with me from Paris some material for that purpose. The consummation of the experiment was, however, delayed until the summer of that year when I finally had the satisfaction *of seeing rotation effected by alternating currents of different phase, and without sliding contacts or commutator*, as I had conceived a year before. It was an exquisite pleasure but not to compare with the delirium of joy following the first revelation.

Among my new friends was the former Mayor of the city, Mr. Bauzin, whom I had already in a measure acquainted with this and other inventions of mine and whose support I endeavored to enlist. He was sincerely devoted to me and put my project before several wealthy persons but, to my mortification, found no response. He wanted to help me in every possible way and the approach of the first of July, 1919, happens to remind me of a form of "assistance" I received from that charming man, which was not financial but none the less appreciated. In 1870, when the Germans invaded the country, Mr. Bauzin had buried a good sized allotment of St. Estèphe of 1801 and he came to the conclusion that he knew no worthier person than myself to consume that precious beverage. This, I may say, is one of the unforgettable incidents to which I have referred. My friend urged me to return to Paris as soon as possible and seek support there. This I was anxious to do but my work and negotiations were protracted owing to all sorts of petty obstacles I encountered so that at times the situation seemed hopeless.

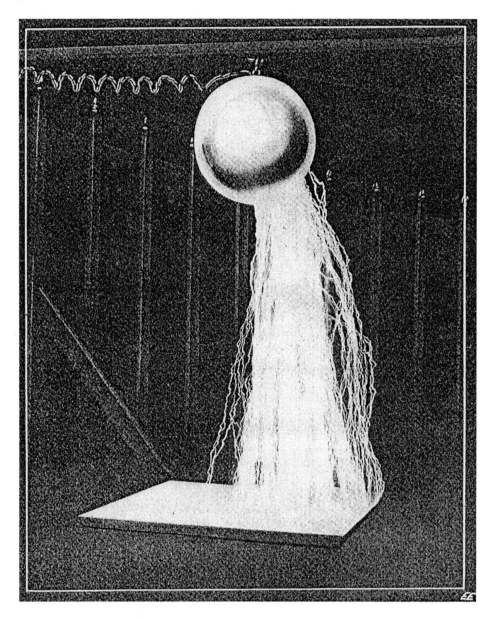

Fig. 2-This Illustrates Tests With Spark Discharges From a Ball of Forty Centimeters Radius in Tesla's Wireless Plant Erected at Colorado Springs in 1899. The Ball Is Connected to the Free End of a Grounded Resonant Circuit Seventeen Meters in Diameter. The Disruptive Potential of a Ball, Is, According to Tesla, in Volts Approximately V = 75,400 r (r Being in Centimeters), That Is, in This Case 75,400 x 40 = 3,016,000 Volts. The Gigantic Tesla Coil Which Produced These Bolts of Thor Was Capable of Furnishing a Current of 1,100 Amperes In the High Tension Secondary. The Primary Coil Had a Diameter of 51 Feet! This Tesla Coil Produced Discharges Which Were the Nearest Approach to Lightning Ever Made by Man.

German "Efficiency"

Just to give an idea of German thoroness and "efficiency," I may mention here a rather funny experience. An incandescent lamp of 16 c.p. was to be placed in a hallway and upon selecting the proper location I ordered the *monteur* to run the wires. After working for a while he concluded that the engineer had to be consulted and this was done. The latter made several objections but ultimately agreed that the lamp should be placed two inches from the spot I had assigned, whereupon the work proceeded. Then the engineer became worried and told me that Inspector Averdeck should be notified. That important person called, investigated, debated, and decided that the lamp should be shifted back two inches, which was the place I had marked. It was not long, however, before Averdeck got cold feet himself and advised me that he had informed *Ober-Inspector* Hieronimus of the matter and that I should await his decision. It was several days before the *Ober-Inspector* was able to free himself of other pressing duties but at last he arrived and a two hour debate followed, when he decided to move the lamp two inches farther. My hopes that this was the final act were shattered when the *Ober-Inspector* returned and said to me: "*Regierungsrath* Funke is so particular that I would not dare to give an order for placing this lamp without his explicit approval." Accordingly arrangements for a visit from that great man were made. We started cleaning up and polishing early in the morning. Everybody brushed up, I put on my gloves and when Funke came with his retinue he was ceremoniously received. After two hours' deliberation he suddenly exclaimed: "I must be going," and pointing to a place on the ceiling, he ordered me to put the lamp there. It was the exact spot which I had originally chosen.

So it went day after day with variations, but I was determined to achieve at whatever cost and in the end my efforts were rewarded. By the spring of 1884 all the differences were adjusted, the plant formally accepted, and I returned to Paris with pleasing anticipations. One of the administrators had promised me a liberal compensation in case I succeeded, as well as a fair consideration of the improvements I had made in their dynamos and I hoped to realize a substantial sum. There were three administrators whom I shall designate as A, B, and C for convenience. When I called on A *he* told me that B had the say. This gentleman thought that only C could decide and the latter was quite sure that A alone had the power to act. After several laps of this *circulus vicious*, it dawned upon me that my reward was a castle in Spain. The utter failure of my attempts to raise capital for development was another disappointment and when Mr. Batchellor pressed me to go to America with a view of redesigning the Edison machines, I determined to try my fortunes in the Land of Golden Promise. But the chance was nearly missed. I liquified my modest assets, secured accommodations and found myself at the railroad station as the train was pulling out. At that moment I discovered that my money and tickets were gone. What to do was the question. Hercules had plenty of time to deliberate but I had to decide while running alongside the train with opposite feelings surging in my brain like condenser oscillations. Resolve, helped by dexterity, won out in the nick of time and upon passing thru the usual experiences, as trivial as unpleasant, I managed to embark for New York with the remnants of my belongings, some poems and articles I had written, and a package of calculations relating to solutions of an unsolvable integral and to my flying machine. During the voyage I sat most of the time at the stern of the ship watching for an opportunity to save somebody from a watery grave, without the slightest thought of danger. Later when I had absorbed some of the practical American sense I shivered at the recollection and marvelled at my former folly.

Tesla in America

I wish that I could put in words my first impressions of this country. In the Arabian Tales I read how genii transported people into a land of dreams to live thru delightful adventures. My case was just the reverse. The genii had carried me from a world of dreams into one of realities. What I had left was beautiful, artistic and fascinating in every way; what I saw here was machined, rough and unattractive. A burly policeman was twirling his stick which looked to me as big as a log. I approached him politely with the request to direct me. "Six blocks down, then to the left," he said, with murder in his eyes. "Is this America?" I asked myself in painful surprise. "It is a century behind Europe in civilization." When I went abroad in 1889--five years having elapsed since my arrival here--I became convinced *that it was more than one hundred years AHEAD of Europe* and nothing has happened to this day to change my opinion.

Tesla Meets Edison

The meeting with Edison was a memorable event in my life. I was amazed at this wonderful man who, without early advantages and scientific training, had accomplished so much. I had studied a dozen languages, delved in literature and art, and had spent my best years in libraries reading all sorts of stuff that fell into my hands, from Newton's "*Principia*" to the novels of Paul de Kock, and felt that most of my life had been squandered. But it did not take long before I recognized that it was the best thing I could have done. Within a few weeks I had won Edison's confidence and it came about in this way.

The S.S. *Oregon*, the fastest passenger steamer at that time, had both of its lighting machines disabled and its sailing was delayed. As the superstructure had been built after their installation it was impossible to remove them from the hold. The predicament was a serious one and Edison was much annoyed. In the evening I took the necessary instruments with me and went aboard the vessel where I stayed for the night. The dynamos were in bad condition, having several short-circuits and breaks, but with the assistance of the crew I succeeded in putting them in good shape. At five o'clock in the morning, when passing along Fifth Avenue on my way to the shop, I met Edison with Batchellor and a few others as they were returning home to retire. "Here is our Parisian running around at night," he said. When I told him that I was coming from the *Oregon* and had repaired both machines, he looked at me in silence and walked away without another word. But when he had gone some distance I heard him remark: "Batchellor, this is a d-n good man," and from that time on I had full freedom in directing the work. For nearly a year my regular hours were from 10:30 A.M. until 5 o'clock the next morning without a day's exception. Edison said to me: "I have had many hard-working assistants but you take the cake." During this period I designed twenty-four different types of standard machines with short cores and of uniform pattern which replaced the old ones. The Manager had promised me fifty thousand dollars on the completion of this task but it turned out to be a practical joke. This gave me a painful shock and I resigned my position.

Immediately thereafter some people approached me with the proposal of forming an arc light company under my name, to which I agreed. Here finally was an opportunity to develop the motor, but when I broached the subject to my new associates they said: "No, we want the arc lamp. We don't care for this alternating current of yours." In 1886 my system of arc lighting was perfected and adopted for factory and municipal lighting, and I was free, but with no other possession than a beautifully engraved certificate of stock of hypothetical value. Then followed a period of struggle in the new medium for which I was not fitted, but the reward came in April, 1887, the Tesla Electric Company was organized, providing a laboratory and facilities. The motors I built there were exactly as I had imagined them. I made no attempt to improve the design, but merely reproduced the pictures as they appeared to my vision and the operation was always as I expected.

In the early part of 1888 an arrangement was made with the Westinghouse Company for the manufacture of the motors on a large scale. But great difficulties had still to be overcome. My system was based on the use of low frequency currents and the Westinghouse experts had adopted 133 cycles with the object of securing advantages in the transformation. They did not want to depart from their standard forms of apparatus and my efforts had to be concentrated upon adapting the motor to these conditions. Another necessity was to produce a motor capable of running efficiently at this frequency on two wires which was not easy of accomplishment.

At the close of 1889, however, my services in Pittsburgh being no longer essential, I returned to New York and resumed experimental work in a laboratory on Grand Street, where I began immediately the design of high frequency machines. The problems of construction in this unexplored field were novel and quite peculiar and I encountered many difficulties. I rejected the inductor type, fearing that it might not yield perfect sine waves which were so important to resonant action. Had it not been for this I could have saved myself a great deal of labor. Another discouraging feature of the high frequency alternator seemed to be the inconstancy of speed which threatened to impose serious limitations to its use. I had already noted in my demonstrations before the American Institution of Electrical Engineers that several times the tune was lost, necessitating readjustment, and did not yet foresee, what I discovered long afterwards, a means of operating a machine of this kind at a speed constant to such a degree as not to vary more than a small fraction of one revolution between the extremes of load.

The Invention of the Tesla Coil

From many other considerations it appeared desirable to invent a simpler device for the production of electric oscillations. In 1856 Lord Kelvin had exposed the theory of the condenser discharge, but no practical application of that important knowledge was made. I saw the possibilities and undertook the development of induction apparatus on this principle. My progress was so rapid as to enable me to exhibit at my lecture in 1891 a coil giving sparks of *five inches*. On that occasion I frankly told the engineers of a defect involved in the transformation by the new method, namely, the loss in the spark gap. Subsequent investigation showed that no matter what medium is employed, be it air, hydrogen, mercury vapor, oil or a stream of electrons, the efficiency is the same. It is a law very much like that governing the conversion of mechanical energy. We may drop a weight from a certain height vertically down or carry it to the lower level along any devious path, it is immaterial insofar as the amount of work is concerned. Fortunately however, this drawback is not fatal as by proper proportioning of the resonant circuits an *efficiency of 85 per cent* is attainable. Since my early announcement of the invention it has come into universal use and wrought a revolution in many departments. But a still greater future awaits it. When in 1900 I obtained powerful discharges of 100 feet and flashed a current around the globe, I was reminded of the first tiny spark I observed in my Grand Street laboratory and was thrilled by sensations akin to those I felt when I discovered the *rotating magnetic field*.

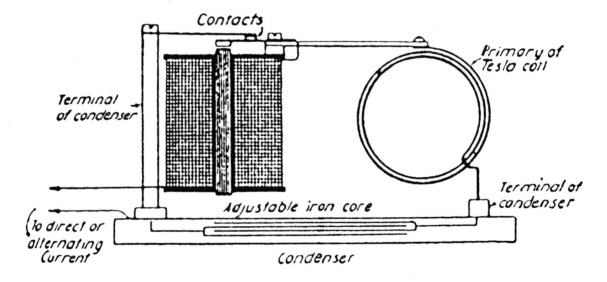

Fig. 3-Scheme of Circuit Connections In Tesla's Oscillation Transformer Shown in Fig.1. The Secondary Circuit Which Slips Into the Primary Is Omitted.

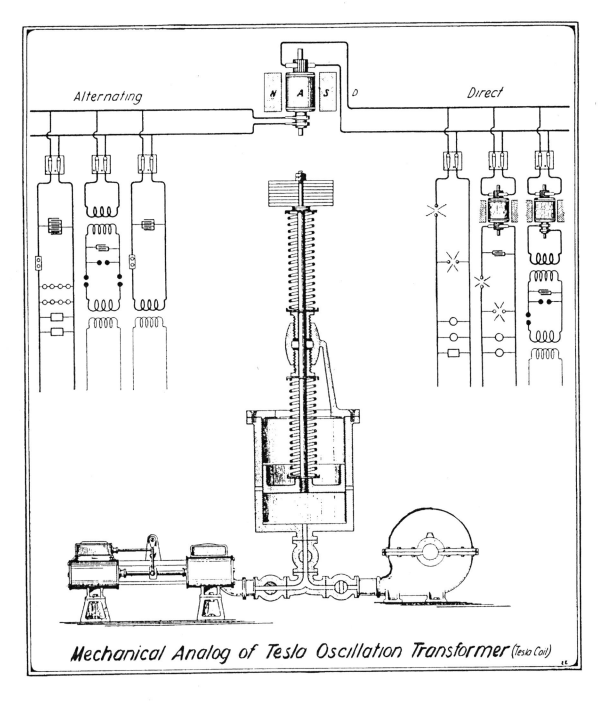

This revolutionary improvement was exhibited and explained by Tesla for the first time in his lecture before the American Institute of Electrical Engineers May 20, 1891. It has made possible to generate automatically damped or undamped oscillation of any desired frequency and what is equally important of perfectly constant period. It has been instrumental in many great achievements and its use has become universal. The underlying principle may be briefly stated as follows: A source of electricity is made to charge a condenser and when the difference of potential at the terminals of the latter has reached a predetermined value, an

air-gap is bridged, permitting the accumulated energy to be discharged through a circuit under resonant conditions, this resulting in a long series of isochronous impulses. These are either directly used or converted to any desired volume or pressure by means of a second circuit inductively linked with the first and tuned to the same. The above diagram is taken from Tesla's lecture before the Franklin Institute and National Electric Light Association in 1893 and shows more elaborate arrangements of circuits, now quite familiar, for the conversion of ordinary direct or alternating currents into high frequency oscillations by this general method. In the mechanical apparatus illustrated, an attempt is made to convey an idea of the electrical operations as closely as practicable. The reciprocating and centrifugal pumps, respectively, represent an alternating and a direct current generator. The water takes the place of the electric fluid. The cylinder with its elastically restrained piston represents the condenser. The inertia of the moving parts corresponds to the self-induction of the electric circuit and the wide ports around the cylinder, through which the fluid can escape, perform the function of the air-gap. The operation of this apparatus will now be readily understood. Suppose first that the water is admitted to the cylinder from the centrifugal pump, this corresponding to the action of a continuous current generator. As the fluid is forced into the cylinder, the piston moves upward until the ports are uncovered, when a great quantity of the fluid rushes out, suddenly reducing the pressure so that the force of the compressed spring asserts itself and sends the piston down, closing the ports, whereupon these operations are repeated in as rapid succession as it may be desired. Each time the system, comprising the piston, rod, weights and adjustable spring receives a blow, it quivers at its own rate which is determined by the inertia of the moving parts and the pliability of the spring exactly as in the electrical system the period of the circuit is determined by the self-induction and capacity. Under the best conditions the natural period of the elastic system will be the same as that of the primarily impressed oscillations, and then the energy of the movement will be greatest. If, instead of the centrifugal, the reciprocating pump is employed, the operation is the same in principle except that the periodic impulses of the pump impose certain limitations. The best results are again obtained when synchronism is maintained between these and the natural oscillations of the system.

Following is the patent of Tesla's most controversial invention, the Apparatus for Transmitting Electrical Energy.

UNITED STATES PATENT OFFICE

NIKOLA TESLA OF NEW YORK, N.Y.

APPARTATUS FOR TRANSMITTING ELECTRICAL ENGERGY

SPECIFICATION forming part of Letters Patent No. 1,119,732.
Patented Dec. 1, 1914.

To all whom it may concern:

Be it known that I, NIKOLA TESLA, a citizen of the United States, residing in the borough of Manhattan, in the city, county, and State of New York, have invented certain new and useful Improvements in Apparatus for Transmitting Electrical Energy, of which the following is a specification, reference being had to the drawing accompanying and forming a part of the same.

In endeavoring to adapt currents or discharges of very high tension to various valuable uses, as the distribution of energy through wire from central plants to distant places of consumption, or the transmission of powerful disturbances to great distances, through the natural or non-artificial media, I have encountered difficulties in confining considerable amounts of electricity to the conductors and preventing its leakage over their supports, or its escape into the ambient air, which always takes place when the electric surface density reaches a certain value.

The intensity of the effect of a transmitting circuit with a free or elevated terminal is proportionate to the quantity of electricity displaced, which is determined by the product of the capacity of the circuit, the pressure, and the frequency of the currents employed. To produce an electrical movement of the required magnitude it is desirable to charge the terminal as highly as possible, for while a great quantity of electricity may also be displaced by a large capacity charged to low pressure, there are disadvantages met with in many cases when the former is made too large. The chief of these are due to the fact that an increase of the capacity entails a lowering of the frequency of the impulses or discharges and a diminution of the energy of vibration. This will be understood when it is borne in mind, that a circuit with a large capacity behaves as a slackspring, whereas one with a small capacity acts like a stiff spring, vibrating more vigorously. Therefore, in order to attain the highest possible frequency, which for certain purposes is advantageous and, apart from that, to develop the greatest energy in such a transmitting circuit, I employ a terminal of relatively small capacity, which I charge to as high a pressure as practicable. To accomplish this result I have found it imperative to so construct the elevated conductor, that its outer surface, on which the electrical charge chiefly accumulates, has itself a large radius of curvature, or is composed of separate elements which, irrespective of their own radius of curvature, are arranged in close proximity to each other and so, that the outside ideal surface enveloping them is of a large radius. Evidently, the smaller the radius of curvature the greater, for a given electric displacement, will be the surface-density and, consequently, the lower the limiting pressure to which the terminal may be charged without electricity escaping into the air. Such a terminal I secure to an insulating support entering more or less into its interior, and I likewise connect the circuit to it inside or, generally, at points where the electric density is small. This plan of constructing and supporting a highly charged conductor I have found to be of great practical importance, and it may be usefully applied in many ways.

Referring to the accompanying drawing, the figure is a view in elevation and part section of an improved free terminal and circuit of large surface with supporting structure and generating apparatus.

The terminal D consists of a suitably shaped metallic frame, in this case a ring of nearly circular cross section, which is covered with half spherical metal plates P P, thus constituting a very large conducting surface, smooth on all places where the electric charge principally accumulates. The frame is carried by a strong platform expressly provided for safety appliances, instruments of observation, etc., which in turn rests on insulating supports F F. These should penetrate far into the hollow space formed by the terminal, and if the electric density at the points where they are bolted to the frame is still considerable, they may be specially protected by conducting hoods as H.

A part of the improvements which form the subject of this specification, the transmitting circuit, in its general features, is identical with that described and claimed in my original Patents Nos. 645,576 and 649,621. The circuit comprises a coil A which is in close inductive relation with a primary C, and one end of which is connected to a ground-plate E, while its other end is led through a separate self-induction coil B and a metallic cylinder B' to the terminal D. The connection to the latter should always be made at, or near the center, in order to secure a symmetrical distribution of the current, as otherwise, when the frequency is very high and the flow of large volume, the performance of the apparatus might be impaired. The primary C may be excited in any desired manner, from a suitable source of currents G, which may be an alternator or condenser, the important requirement being that the resonant condition is established, that is to say, that the terminal D is charged to the maximum pressure developed in the circuit, as I have specified in my original patents before referred to. The adjustments should be made with particular care when the transmitter is one of great power, not only on account of economy, but also in order to avoid danger. I have shown that it is practicable to produce in a resonating circuit as E A B B' D immense electrical activities, measured by tens and even hundreds of thousands of horse-power, and in such a case, if the points of maximum pressure should be shifted below the terminal D, along coil B, a ball of fire might break out and destroy the support F or anything else in the way. For the better appreciation of the nature of this danger it should be stated, that the destructive action may take place with inconceivable violence. This will cease to be surprising when it is borne in mind, that the entire energy accumulated in the excited circuit, instead of requiring, as under normal working conditions, one quarter of the period or more for its transformation from static to kinetic form, may spend itself in an incomparably smaller interval of time, at a rate of many millions of horse power. The accident is apt to occur when, the transmitting circuit being strongly excited, the impressed oscillations upon it are caused, in any manner more or less sudden, to be more rapid than the free oscillations. It is therefore advisable to begin the adjustments with feeble and somewhat slower impressed oscillations, strengthening and quickening them gradually, until the apparatus has been brought under perfect control. To increase the safety, I provide a convenient place, preferably on terminal D, one or more elements or plates either of somewhat smaller radius of curvature or protruding more or less beyond the others (in which case they may be of larger radius of curvature) so that, should the pressure rise to a value, beyond which it is not desired to go, the powerful discharge may dart out there and lose itself harmlessly in the air. Such a plate, performing a function similar to that of a safety valve on a high pressure reservoir, is indicated at V.

Still further extending the principles underlying my invention, special reference is made to coil B and conductor B'. The latter is in the form of a cylinder with smooth or polished surface of a radius much larger than that of the half spherical elements P P, and widens out at the bottom into a hood H, which should be slotted to avoid loss by eddy currents and the purpose of which will be clear from the foregoing. The coil B is wound on a frame or drum D' of insulating material, with its turns close together. I have discovered that when so wound the effect of the small radius of curvature of the wire itself is overcome and the coil behaves as a conductor of large radius of curvature, corresponding to that of the drum. This feature is of considerable practical importance and is applicable not only in this special instance, but generally. For example, such plates at P P of terminal D, though preferably of large radius of curvature, need not be necessarily so, for provided only that the individual plates or elements of a high potential conductor or terminal are arranged in proximity to each other and with their outer boundaries along an ideal symmetrical enveloping surface of a large radius of curvature, the advantages of the invention will be more or less fully realized. The lower end of the coil B--which, if desired, may be extended up to the terminal D--should be somewhat below the uppermost turn of coil A. This, I find lessens the tendency of the charge to break out from the wire connecting both and to pass along the support F'.

Having described my invention, I claim:

1. As a means for producing great electrical activities a resonant circuit having its outer conducting boundaries, which are charged to a high potential, arranged in surfaces of large radii of curvature so as to prevent leakage of the oscillating charge, substantially as set forth.

2. In apparatus for the transmission of electrical energy a circuit connected to ground and to an elevated terminal and having its outer conducting boundaries, which are subject to high tension, arranged in surfaces of large radii of curvature substantially as, and for the purpose described.

3. In a plant for the transmission of electrical energy without wires, in combination with a primary or exciting circuit a secondary connected to ground and to an elevated terminal and having its outer conducting boundaries, which are charged to a high potential, arranged in surfaces of large radii of curvature for the purpose of preventing leakage and loss of energy, substantially as set forth.

4. As a means for transmitting electrical energy to a distance through the natural media a grounded resonant circuit, comprising a part upon which oscillations are impressed and another for raising the tension, having its outer conducting boundaries on which a high tension charge accumulates arranged in surfaces of large radii of curvature, substantially as described.

5. The means for producing excessive electric potentials consisting of a primary exciting circuit and a resonant secondary having its outer conducting elements which arc subject to high tension arranged in proximity to each other and in surfaces of large radii of curvature so as to prevent leakage of the charge and attendant lowering of potential, substantially as described.

6. A circuit comprising a part upon which oscillations are impressed and another part for raising the tension by resonance, the latter part being supported on places of low electric density and having its outermost conducting boundaries arranged in surfaces of large radii of curvature, as set forth.

7. In apparatus for the transmission of electrical energy without wires a grounded circuit the outer conducting elements of which have a great aggregate area and are arranged in surfaces of large radii of curvature so as to permit the storing of a high charge at a small electric density and prevent loss through leakage, substantially as described.

8. A wireless transmitter comprising in combination a source of oscillations as a condenser, a primary exciting circuit and a secondary grounded and elevated conductor the outer conducting boundaries of which are in proximity to each other and arranged in surfaces of large radii of curvature, substantially as described.

9. In apparatus for the transmission of electrical energy without wires an elevated conductor or antenna having its outer high potential conducting or capacity elements arranged in proximity to each other and in surfaces of large radii of curvature so as to overcome the effect of the small radius of curvature of the individual elements and leakage of the charge, as set forth.

10. A grounded resonant transmitting circuit having its outer conducting boundaries arranged in surfaces of large radii of curvature in combination with an elevated terminal of great surface supported at points of low electric density, substantially as described.

NIKOLA TESLA.

Witnesses:
    M. Lamson Dyer,
    Richard Donovan.

N. TESLA
APPARATUS FOR TRANSMITTING ELECTRICAL ENERGY

No. 1,119,732.   Patented Dec. 1, 1914.

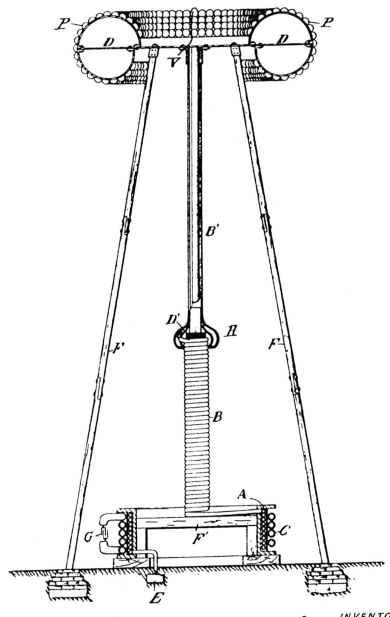

ELIHU THOMSON was born in Manchester, England in 1853 and died in Swampscott, Massachusetts in 1937. His inventions helped develop the alternating current motor industry. He created some 700 patents. Some of these were in electric welding, X-ray tubes, watt-hour meters, alternating current motors, high frequency generators, and high frequency transformers. The company he co-founded, the Thomson-Houston Electric Company merged with the Edison General Electric Company to form the present General Electric Company.

Following is an address given by Thomson in 1890. It was reprinted in a booklet; Thomson, E., Transactions of the Thomson Scientific Club, "What is Electricity?", (Lynn, Mass.: Burbier Publishing Co., 1904), p.1-30.

"WHAT IS ELECTRICITY?"

IN Nature we have a vast exhibition of activity and active agents. We recognize certain actions or phenomena as belonging to one class and other actions or phenomena to a different class. In this way we study effects of gravitating force, waves of sound, the action of heat and light, we study the transmutation brought about by chemical forces, and lastly we have to deal with a great variety of actions and effects which we call electric and magnetic.

I have taken the subject "What is Electricity?" for this evening's talk. According to the advertisements of quack apparatus, called electric belts, brushes, corsets, and, most interesting to what was a shoe town before the electric factor came, electric insoles, the question we have proposed for ourselves is easy to answer--"Electricity is Life," and "it cures all diseases, none genuine but the electricity *our* belt or *our* insole or what not, makes"- and this may be quoted substantially from numerous such advertisements. The New York troubles over electric wires must have done something to disabuse the public mind of belief in such definition; unless having too much life is injurious or fatal. I remember one such so called electric battery or amulet which was evidently constructed of the scrap brass and other metal of some metal works. It contained about two cents worth of metal and sold for fifty cents. Its peculiar virtue was as stated in the advertisements; that it made the electricity in a gimlet form which penetrated the vital organs. The same advertisement said that electricity was life, so that life in a gimlet form must have had a bad effect, for the thing quietly dropped out of sight, ceased to be exchangeable for a piece of silver of the same size, and was forgotten in less time than is taken by various new nickel-in-the-slot catching apparatus, in sinking into oblivion.

But to return to our question, What is electricity? Can we answer it? No--we, can only point out the lines of progress in that direction and hope that sometime in the future the secret will be out, at least so far as finite mind can obtain an answer.

Not so very long ago--in fact, not much over a hundred years ago, anything rather light or intangible or mysterious was explained by, assuming an essence or spirit, a sort refined emanation or effluvium. Hence when wine was distilled there was obtained the lighter, more volatile product spirit of wine, or our alcohol. With turpentine or pine sap distilled it was spirit of turpentine, while muriatic acid was spirit of salt, being made from salt, and sweet spirits of nitre was supposed to come out of nitre or nitrate of potash distilled with alcohol and sulphuric acid. Electricity likewise was a refined spirit of matter, an effluvium. Naturally in his beginnings with the sciences, the reasoning of man was tinged with preconceived notions of what ought to be, and he was often very forcibly reminded that he must not get outside of certain limits of thought. True science knows nothing of what ought to be but only endeavors to use what faculties we possess to discover what is.

Cotton Mather, an active spirit in condemnation of witches in New England, was doubtless convinced that they ought to exist if they, did not, just as he was convinced that thunder and lightning was due to the explosions of sulphurous and nitrous components of the atmosphere, and, as I am told by one who studied the matter; was convinced that he knew enough to be a member of the Royal Society if he was not, and so printed on the title pages of his books, "By Cotton Mather, Member of the Royal Society." I allude to these matters incidentally as showing the different attitude which existed only a few decades ago towards truth than exists to-day.

The strength of science is in ascertained and proved facts. We have learned to investigate, to approach the fountain of truth with a true humility; to doubt the correctness of preconceived ideas; to be always ready for new truths which can be proven; to be happy in our research, to feel that the moral welfare and condition of humanity depends upon a knowledge and mastery of the physical conditions of existence. We have learned to regard the wonderful play of energies in this world, and in the mighty depths of the Universe, with a truly religious enthusiasm and awe. The more deeply we penetrate the veil which hides both the mysteries and the beauties of the physical universe, the more are we impressed with a sense of our own insignificance, of our own powerlessness to reach a full understanding of its complex relations, and of the abundant field for the future acquisition of knowledge.

Perhaps no branch of science has yielded so much to the searcher as electricity has done in the past few years. Scarcely more than a century has passed since its simpler actions, almost its existence except in lightning, have been known, scarcely more than a half century has passed since it was called upon to do any useful work, as in telegraphy. And what have we at present? An industrial giant, a many branched tree, an organism as it were whose proportions are only rivalled by the variety and refinement of its organs; the telephone, with its inexpressibly delicate currents, the electric light, the electric motor and railway, the electric metal working industry which, besides telegraphy, are each separate arts in themselves.

I ought not to leave out of the list electro-metallurgy, one of the oldest of them all, but which is now applied to the production of aluminium, the extraction and purification of metals, and the plating of metals generally. There are many other things that electricity can do, and it needs only time and suitable appliances to work them out.

Now, let us endeavor to see what this great force is, as well as we can. Formerly it was thought to be an imponderable fluid, then two such fluids, positive and negative.

The simple hypothesis of Franklin explained electrical action by simply saying that, when a body was excited or electrified positively, it was in possession of a little more or much more electricity than is natural or usual. When it was excited negatively it was in possession of less than its natural share. He simply called electricity a fluid and then said that it flowed along a wire from positive to negative. This view, however, does not explain all the effects. It leaves us, in the dark just as much as before. With this hypothesis came the hypothesis that electricity was two fluids which, when they were brought together or near together, showed a strong tendency to unite, and after there was actual contact between two conductors having each a charge, one positive and the other negative, the two would run together and produce equilibrium. That view was held for some time and used as an explanation. There are undoubtly two stages of electrical action, though that there are two electricities differing from each other in many different ways is a little difficult, I think to conceive. But after all if we call electricity, a fluid then we come back to the question, what is a fluid? It is easy enough to explain a thing by saying that it is something you do not understand. We say that water is composed of oxygen and hydrogen. Yet there is something else with them. When we take these two gases and mix them in the proper proportion and fire the mixture we get an explosion. We get a particle of water, and neglect the explosion. But there was certainly energy given out besides. There was an energy which came from the oxygen and hydrogen with the result of an explosion, and the ashes, or thing left as it were, was the water. Now, we come back, however, to what we started with--what is oxygen? What is hydrogen? We say that they are chemical elements, but this leads us only a little furthur and we ask what are the things we call elements? What is matter anyhow? There we are puzzled. We stop right there. We have reached, our limit unless we go still further and say it is composed of little atoms, little fine particles, and we make them so small that we cannot see them. Then we have got rid of our difficulty. (Laughter.) But we will imagine that we magnify these small atoms so that they can be seen. We come to the same problem as before. If the atom has something inside it, we ought to break it up with infinite force, and we ought to be able to see what is inside of it, or what composes it. This puzzling problem is really at the foundation of Nature, of the constitution of the universe. If we accept the fluid idea of electricity we have to explain what a fluid is, which leaves us in just as much difficulty as before. Heat at one time was regarded as a fluid. It was regarded as a sort of thin fluid which came out of hot bodies. When a bar of iron was put into the fire and was made hot and held up so that the hand could feel the heat, it was supposed that there was a fluid streaming out from the bar, an imponderable fluid, because the bar having lost its heat did not weigh any less. In the same manner the first theory of light was that it was fluid, it was a fluid which could be caught in a lens and brought to a focus so as to set fire to things, as in a burning glass. So we have heat, light and electricity all explained by saying that they were so many different fluids.

The object of my lecture this evening will be to show that we can go a little distance in proving the identity of light with electricity. The actual experimental proofs of this have been developed only within the last year or two. It was Faraday, who, for the first time, seemed to have the idea pretty clearly, although others undoubtedly had glimmerings of the same idea. Clerk Maxwell, an eminent mathematician and physicist worked out this fact. He worked at the problem mathematically and came to the conclusion that the vibrations of light were electric waves, that what we call undulations of light were electric waves. I should say here that prior to this the fluid idea of heat had been abandoned and other ideas more in accordance with the facts were held. Heat was not a thing. It was the result of a movement, of vibrations of the finer portions of matter, a vibration of the small masses or molecules as we call them. When a body is heated, when this bar which I hold in my hand, for instance, is heated its particles are shaking more than they do now. When I say that this bar is hot after putting it into the fire it means simply that its particles are violently shaking more than they do now. If I put it into a freezing mixture I make it vibrate less. There the swinging is less. When I make it hot the particles move through greater distances. The idea of calling light a fluid was very soon changed in the same way as occurred with heat. It came to be regarded as undulation, but the question always arose then--undulation of what? It is perfectly certain that if we exhaust the air, take as much air as we can out of a vessel by the best air pumps, the passage of light is not hindered in the least. If I take this bulb of an incandescent lamp and examine it, I find that there is a very small amount of the original air remaining in it. It is almost a perfect vacuum and yet you see that the passage of light through it has not been affected in any way. The space carries the light with as perfect freedom as the air itself. Consequently, if light is vibration of something, what is that something? It is assumed, and has had to be assumed, that there exists what is known as the universal ether; an ether which fills all space--a light, imponderable, almost intangible, medium, whatever else it may be. So that after all we still come back to a fluid, a fluid that fills all space. We will have succeeded in doing a great deal if we have only one fluid to find out the nature of instead of a great many.

To elucidate some of the effects we call vibrations I may call attention simply to the fact that in this open organ pipe we have a body of air contained and by blowing into the end

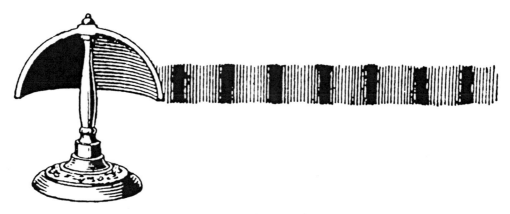

FIG. 1.

(illustrating) we may set that body of air into vibration. The actual manner of motion is this. The air divides itself into two masses, apparently stationary at the middle of the pipe body itself, and moving masses nearer the ends, that is, the column is composed of portions which are not in motion and then portions which are vibrating. If I blow into this pipe the air that comes out of the lip of the pipe here strikes this upper edge and sets up a vibration in the air. That is a well known phenomenon. This principle is used in making organs. Now, how do we become acquainted with the effects of this vibration? We hear it you say, but it was a long time before even sound was known to be a vibration. It was thought at one time to be a sort of effluvium, shaken out of sounding bodies as it were. We now know that it is nothing more nor less than a kind of motion in the air propagated at the rate of 1000 feet in every second, a rate which can be easily exceeded by a rifle bullet, the bullet striking the target before the sound gets there. We call that nowadays a rather slow motion. We will look at the manner of propagation for a moment. (Blackboard) Here we have a bell sending out sound waves, (Fig. 1). We may have first the air pushed strongly all around the bell and made dense. Then we find that this dense wave moves along, instead of remaining close to the bell it is thrown off or proceeds onward. Then the bell moves back and makes a weaker or rarified portion next <to> it. The air at this point, is then not so dense. After the sound has gone a distance, we find that these dense and rare portions alternate with each other. The bell strikes the air about it, and sets it into vibration. This is the longitudinal propagation of waves. The motion is that way,

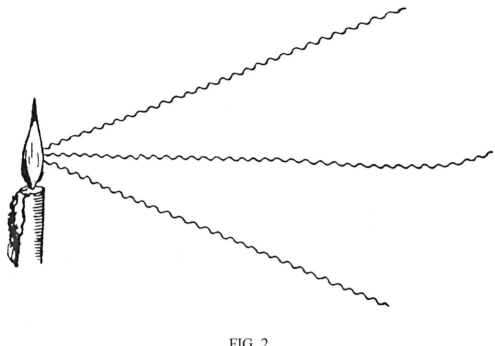

FIG. 2.

(indicating), right straight ahead, a backward and forward motion of the air particles. It is similar to the vibration which can be produced in other objects, such as solid rods, by striking them on end the sound wave being propagated to the other end. This steel bar will vibrate if I strike it on end.

In light the vibrations do not occur in that way. They are not in air but in the ether. In light the vibrations are in a direction transverse to those of sound. The particles of the ether surrounding the luminous body, the particles, so to speak, of that universal medium which fills all space, are set to vibrating transversely, to the direction of the ray, (Fig. 2). This has been determined mathematically, and by a great number of optical experiments. It has been difficult to understand why it should vibrate in transverse vibrations. A good deal of light, however, has been thrown upon the subject.

Now, when we come to deal with electricity we find that the phenomena are very various. There seems to be a greater variety in them than in the case of heat and light, and it is not surprising when we really study into the question as far as we can. If I rub this piece of black vulcanite or hard rubber, it becomes electrified as it is termed. This is the first form in which electrical action was known. The rubbing of a piece of amber developed electrification. It was then known to attract or repel light bodies. There is a slight crackling sound produced during the rubbing. I hold the rubber up to these light bodies suspended here and we see that the first effect is attraction. Now touch both of them and they are not attracted any more--they

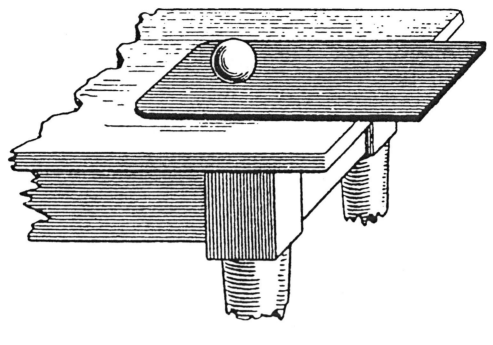

FIG. 3.

are repelled unless I get very near and force them to take more electricity. Now I cannot get near them at all. Here, then is a condition which is quite surprising, and must have been a great surprise to those who first investigated it. I let them touch my finger and they will come together as at first.

Still another experiment of the same nature and one showing another phase of this electrical action is to take a light object, a piece of sunflower pith, (the pith of the stalk of sunflower dried makes a very excellent substance for working in this way.) I will take that rubber and put it over the piece of pith on the table and you will see that there is an attraction of the pith. Now, you see the pith jumping toward the rubber and now a repulsion and a jumping again as though there was something that the pith was getting from the rubber and carrying down to the metal plate. But curiously enough, if I lay this rubber on the table and vigorously rub it, it seems inactive for I can put the pith on it without repulsion, (Fig. 3). If I wish to keep it on that rubber I must lay the rubber down. Now, the rubber is lying on the table. I will rub it vigorously while it is lying flat on the table. Now let us see if we can get any electrical action. Apparently not, the pith is not affected in the least. There it lies just as ordinarily. Suppose I attempt to raise the rubber plate off the table--see, the pith jumps off. We lay it down on the table again and we will be able to get the pith to stay. Raise it again--now it seems to stay. Perhaps we have lost the electrical charge. I will rub it once more. Now I will put it down and we will see what occurs. There, the pith jumps off to this side. The moment I raise the rubber and get it a little distance away from the table all at once the rubber seems to get back into the condition that it was in before. I put light bodies on it just as easily as I please while it rests on the table, but if I raise the rubber they are driven off. I have a little ball here, a little round white ball. There are also three little pieces of elder pith. I will put them all on and I will raise the rubber--they have all gone too. We call that quiescent state another condition of electricity. It is there in a bound state, that is it is bound down to the table and cannot act on the pith. It is exerting itself towards the table. It is occupied continually and will not allow you, as it were, to distract its attention from the table. It is trying to get through and down to the ground, but it finds the rubber too thick, but still it is trying and going to keep trying to get through.

There is another phase of electric action. We may take this electrification and, by means of conductors or wires, we can convey it, or we may rub this piece of rubber and develop the electric charge and we can conduct it away, but to do that we must have conducting materials. The reason that the rubber does not lose its electricity when I hold it in my hand is that the electricity can not run down on account of the rubber not being a conductor. It is a non-conductor and a very excellent non-conductor. If it were metal and electrified the charge would disappear into the floor. It would be gone in so small a fraction of a second as to be almost inconceivable. Now when we have a large amount of this conductive action over a wire we have what is called an electric current. We may have a current of very small amount or of very large amount. It may be so small, as in the telephone, that we can hardly detect it unless we have very sensitive instruments, or we may have currents of enormous volume, of enormous power in so far as they represent a flow of something, which is capable of heating large pieces of steel or iron.

We must give one caution here. Do not think that a large current means that it is a dangerous current, not by any means. The popular idea is that the larger the current passing the more dangerous it must be. That is not the question at all. It is a question of whether the current has sufficient pressure to pass itself through the animal body. There may be danger if the pressure is high enough. In the operation of motors, for example, the moderate pressure is not able to send current through the animal body. You may have millions of horse power passing over a conductor and the current be perfectly harmless even if the conductor is touched so as to complete a circuit through the body. It is a question of pressure sufficient and volume more than sufficient to pass through the animal body to do harm.

We have mentioned several forms or manifestations of electricity. I am not going to mention many more only one more, although I might go on and enumerate many other kindred phenomena. I shall now call attention to another manifestation which for a long time was not believed to be electrical, yet it has always accompanied electrical action, and that is the action of magnetism. Here we have something, which, instead of residing in an insulator when rubbed like that (illustrating), may reside in a piece of steel that has been hardened, and a piece of steel which has been touched by another piece of steel in a magnetic condition. It is what we call a magnet. This is a different manifestation from any of the others. Whenever we have electricity moving from one point to another, we always have this other manifestation of magnetism. If then we make a circuit, as we call it--if we take a small battery jar and put in it a chemical fluid and two plates of different metals, and then make a circuit (Fig. 4), no matter how long or short, we will have current flow through that circuit. We have around the wire at all times a magnetic field, as we term it. The current magnetizes, not air because if we take away the air you will find the magnetism there still. Well, then it must magnetize the universal ether which is present everywhere, whether the air is present or not. This ether is present all through the solid earth, all through the universe. The current produces lines of strain in it-- magnetic strain. We can take a great many turns of the circuit wire and then we have a coil supplied by current from the battery, or the current may be supplied from some other source of electricity. We will now take a piece of iron and put it into the coil and instantly we find that

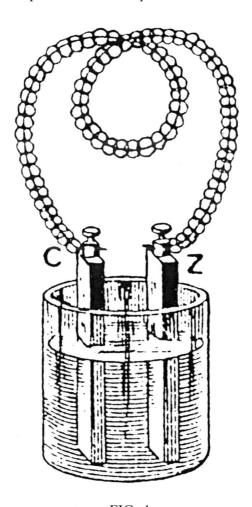

FIG. 4.

the magnetic effect is enormously enhanced. What would be produced in the ether alone is now magnified hundreds and even thousands of times. We have found something which is better than ether, we have found, as it were, a better atmosphere. The iron is a better atmosphere, replacing the ether, so that it can accept a great many more strain lines, as we may call them, of magnetism, a great deal more of this magnetic strain.

It would take too long to enter into a discussion of this department of electricity. We would have the whole science in fact to ransack and it would take evening after evening to begin to give you the elements of it. My object on this occasion is to call attention to some phenomena which attract attention to the nature of electricity. They do not tell us what it is, but tell us it is like something else. Recent developments in the industrial world have brought to our notice certain kinds of current, electric currents known as alternating currents, currents which are not like those of the battery current, all the time going in the circuit in the same direction, but which go around one way and very soon after go around in the other way, and then reverse again. They may alternate very slowly or they may alternate very rapidly. The current which we are about to use in the illustrations here and coming in by these wires alternates at the rate of 250 to 260 times per second. Notwithstanding their rapidity these electric impulses get all around Lynn and back to the dynamo easily enough before the next impulse starts out in the opposite direction. You know how far these circuits extend, out into the fields, and yet that current goes around and back and goes 250 times in every second. But that is nothing. If we should choose we could make them go around tens of thousands of times in a second, 50,000, 100,000 per second. We would not have much difficulty in doing that. The current would get around and be on time. (Applause). I should not say exactly on time,-- according to the conditions there is a little lag, as we call it, but still it would keep up its rate. Its rate of getting around and back would be all right.

Now when we come to consider the vibrations which make up light, that is, to discover what luminous vibrations are, we find that we are dealing with numbers which are simply stupendous. The ether which fills all space and carries the light from the sun to our earth, carries also radiant heat. Light and radiant heat are the same. When you sit in front of the fire you feel the warmth on your face, that is the same thing as light. You call it heat, but it is radiation which is invisible. Now, I was saying a little while ago that this vibration which leaves the sun, leaves the arc light, gas, or any other light, is composed of transverse vibrations, vibrations which are across the path of the ray. What is the rate at which they move? They go from the sun to a point 200,000 miles away in one second. A distance of 200,000 miles is traversed in one second. But how many waves are there in that 200,000 miles? Is there one wave and then another? No. The reversal of motion here takes place at a very high rate of 500,000,000,000,000 or more in every second. (Writing it out on the blackboard). That is the number. There are then in 200,000 miles of a light wave, in other words, between us and a point this side of the moon, 500,000,000,000,000 of vibrations passing. How many would this make to the inch? We might ask. About 40,000. That is, there are about 40,000 of these little shakes to every inch of progress, and yet it is difficult to conceive that there can be between us and that gas light 40,000 to 60,000 or more shakes to the inch according to the character of the light. Dark red rays give a less number than violet, and the number increases progressively as we go from the red towards the violet end of the spectrum. Some of them can only be photographed, as they are too fine or too fast for our eyes. Our eyes were not made to see them. They may be illustrated by an analogy in sound. If I take that bar and hit it on end it gives out a sound due to what are called longitudinal vibrations. If I should break the bar in half and then hit it, the note would be an octave higher. If I break it again, an octave higher still, until with a very small piece of steel, I would hear nothing but the blow. I should not hear the high ringing sound. It would have gone beyond the limits of hearing, vibrating faster and faster, beyond 40,000 vibrations per second when I could not hear the sound. A little below that, one might hear it.

Now to come back to our alternating current. We produce 250 impulses or 125 complete waves in a second, not much compared with light, and yet the light wave is nothing more nor less than that kind of a wave, and this is what has been proven. If I could take this current and make it go at the rate of 500,000,000,000,000 waves or vibrations in a second, that coil of wire would glow with light. It would give out in the surrounding media waves of light. Now it is giving out impulses or wave motions at the rate of 250 per second. It is giving out waves just the same as light but slower. It gives us waves of magnetism, and the magnetic waves which it gives out, if speeded to 500,000,000,000,000 times per second are light. Light is electro magnetic waves. They are waves in the ether, and they are electric waves. Light is therefore an electric phenomenon. When you are studying optics you are studying electricity. The whole science of optics is an electrical study. It has become a part of the science of electricity. And how was this brought about? Why, by endeavoring to get electric waves of very high rate, by endeavoring to get some of those alternations of current so fast that they would approach the rate of light waves. Investigators have been endeavoring to increase the rate at which the waves could be made to go backward and forward, waves of current to produce waves of magnetism which would be fast enough to be something like light. Now, how near can we get? I said there were 40,000 to the inch in red light. Dr. Hertz succeeded in getting waves which were 3 1/2 feet long. By comparison of these two lengths, 1/40,000 of an inch to 3 1/2 feet, we find how far we have got to go before we get light.

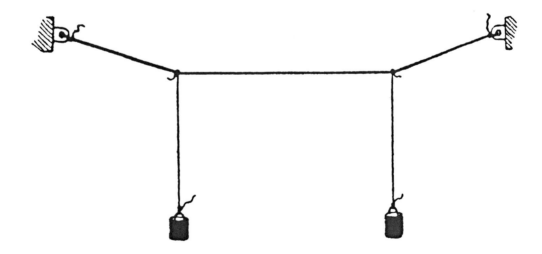

FIG. 5.

Still how do we know that when we get that far, get waves fast enough to be like light in rate, they would be light? Well, there are certain phenomena that are known in sound. These are known in heat, and they are known in light. These are the phenomena of interference, besides which we must notice the phenomena of resonance. In mechanics these phenomena are perfectly well known also. If I take this pendulum and start it moving (two pendulums of equal lengths were supported from a string at equidistant points from the ends, Fig. 5), it does not allow that pendulum to remain still but puts it in motion. (Experiment tried. One pendulum was in a state of rest; the other being swung, its motion was gradually retarded until it stopped, while the first taking up the motion was put into a state of vibration corresponding to the latter. After an interval of time had elapsed the first pendulum stopped gradually and the latter was again in a state of vibration. The lecturer then took up an organ pipe and a tuning fork, striking the tuning fork). You hear the note. The pipe has very nearly the same note, but not quite, and I will get it down to the right tone with my hand. (Striking the fork and holding it over the mouth of the pipe, Fig. 6.) Now the pipe speaks. It is not the fork. It is the pipe. The fork is in harmony with the pipe. The fork tends to set the air in the pipe in motion. The pipe speaks. That is what we call resonance. (The lecturer then went to the piano making the note on the organ pipe). I put my foot on the loud pedal of the piano. You hear the same note on the piano. (Applause). There is a sympathy between a string in the piano and this pipe. Further I strike the tuning fork and hold it over that jar (Fig. 7 A); (a glass jar was on the table); you hear nothing, but I take this water, which I have previously measured and found to be right, and pour it into the jar. We must be careful to get the quantity just right. We can not depart a quarter of an ounce either way. (The fork was struck and held over the jar) Now, you hear it. It is very loud. You hear that the jar sympathizes with the fork

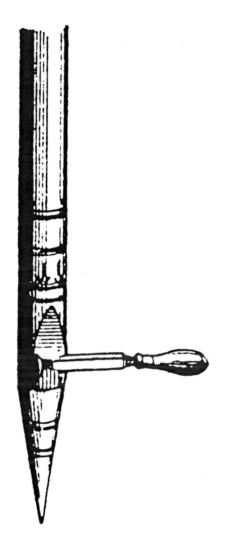

FIG. 6.

as the pipe did. (Applause). (The experimenter then took up an ordinary glass globe, Fig. 8). This is an ordinary glass globe, such as is commonly in use. We will bring its larger opening down into the water in this pail. We will try it with the fork applied to its open end above water with a very little water in the globe. We will get the effect when we get down further. There is a point where (raising and lowering the globe) you will hear the sound to be very pronounced. If I drop it a little too far it will not be in sympathy, and if I raise it a little too far it will not be in sympathy. When the column of air in the globe is of the right volume and shape, or when it is in tune with the fork in other words, then we get this resonance or sympathy. Now we can do this same thing with light. We can multiply such experiments almost endlessly. I only notice them now as showing what has been done in light.

There is another thing that is shown by this experiment. Notice. If I put the fork over the open jar in a special way we will not get the sound. (The lecturer put the tuning fork over the jar with water, with an angular or inclined position of the legs of the fork). That is another phenomenon. It is the phenomenon of interference. There are two legs to this fork. If I turn them so (angling Fig. 7 B), the vibrations of this leg (the lower) reach the column of air in the jar at a different time from those of this leg (the upper) and they interfere. A knowledge of these phenomena show them to be the same as those of light. I cannot go into a full

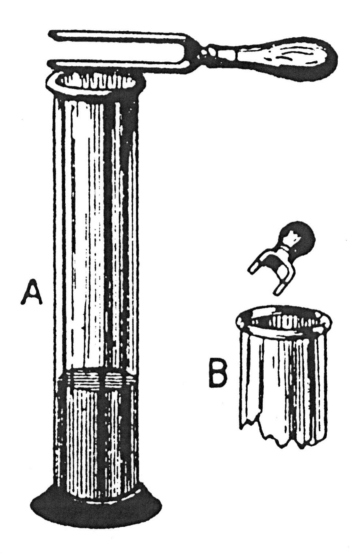

FIG. 7.

explanation. In fact it is almost too far to go for an audience which has not followed the finer developments of optical science, but Dr. Hertz took waves of a kind produced by electric currents, by alternating currents, and he caused them to do just what waves of this kind will do, that is he caused them to interfere with each other. He found places where the waves would wipe each other out. He knew that light waves themselves did just the same thing. He found furthermore, and this is curiously true, that he could take magnetic waves and produce other similar effects. Suppose we take a brilliantly polished parabolic mirror, say of silver, and

FIG. 8.

we put a little coil here and send currents of these very high rates of vibration through the coil (Fig. 9). The electric wire conveying these alternating currents sends electro-magnetic waves out into the space around it and they are reflected from the mirror, according to the laws of light, so as to become a parallel bundle of rays. He then placed at some distance away and in front of the reflector a prism of pitch. He used pitch because it is cheap and because it is an

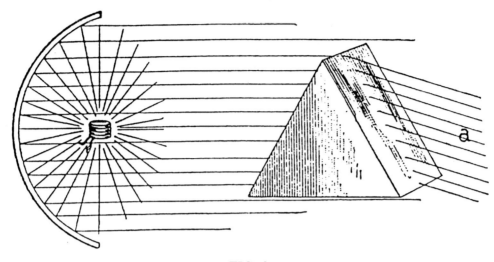

FIG. 9.

insulator. He passed the waves through that prism and they went right through the black pitch--went clear through, but they did not come out in the straight direction. They are turned. They are refracted. They are bent, just as the light was bent, but these waves are 3 1/2 feet long and the light waves are 1/40,000 of an inch long, or less.

How does he know that the waves were refracted in that way? He simply constructs an electrical device like that, (holding up a coil of insulated wire with the ends approached quite near each other.) He constructs an electrical sympathizer, an electrical resonator, and he feels around in the space (moving the coil about a coil of wire wound upon an iron core standing on one end upon the table) with this resonator. He constructs a device which will respond to waves 3 1/2 feet long. He feels around the pitch prism and in the space beyond and just as soon as he gets here in the path of the waves he sees he is in the path of the rays, for the coil gives a little spark between its ends. He now gets further away and feels in this space again (indicating on black-board.) He follows these rays to the point where they leave the prism out through different angular positions, traces them out through space. He has not got eyes that can see waves 3 1/2 feet long, but by the electrical device he can give himself eyes, and that electrical device is nothing more nor less than the tiny coil resonator. (By approaching a coil on the table the lecturer obtained an electric spark which jumped across the open space between the terminals of a coil in his hands. This spark was reduced in intensity upon receding from the core upon the table until at last it was just visible to the lecturer at a distance of about 2 feet from the table.) If he makes that coil too big to respond or sympathize with waves 3 1/2 feet long--he sees nothing; but make it just right--the right length, tune it up, and if there are any of those waves in the space then that little coil will sympathize with them and there will be seen a little spark at its ends. If I had two large coils on opposite sides of this room with the same number of turns, and with Lyden jars of the same capacity connected to them so that they were matched, sparks would be given out from one when the other had a discharge passed through.

I have a number of experiments to show this evening concerning actions of electric waves. I would say before I begin that many of them are novel experiments, some of them being only a week or two old, and others were shown for the first time in public at the Paris exposition last year. I think you will agree with me, when you see the experiments, that there is a something moving away from this coil and capable of producing quite a number of actions. I have a coil of wire here (pointing to the magnetized core referred to above) which is simply a bobbin, and in its interior is placed a bundle of iron wires called a core. When I send a current through this coil it magnetizes the iron. The iron, however, is not alone the seat of this magnetism but it sends lines of magnetism out into the air. The coil only uses the iron as a

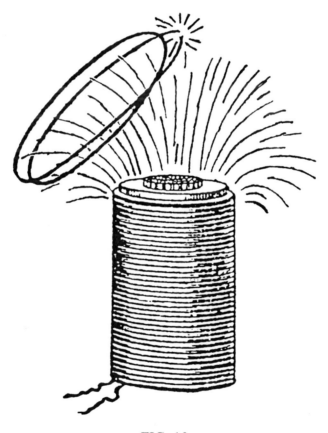

FIG. 10.

better medium, as a means of setting up or conveying magnetism easily. If we send a current through this coil, which is a steady current, it does not send out waves of magnetism, but only becomes magnetized. It is magnetized but does not give out any pulsations. If I send an alternating current through this coil changing magnetism or waves will be given out. I will have to be satisfied this evening with 250 waves per second. We have not here waves which are 3 1/2 feet or even 50 feet in wave length, but a complete wave here has a length of 1600 miles.

We will first prove that there is something going out from this coil. How can, we do that? You hear the humming sound which indicates the vibrations, that is the current going through the coil. (The lecturer then took up a large coil of insulated wire and showed sparks between its ends produced when it was brought in proximity to the magnetized core on the table, Fig. 10). We will next prove that these waves are magnetic waves and capable of affecting iron as a magnet. (The lecturer placed a sheet of glass over the coil and placed a set screw on it. He then passed an iron rod around the screw which was set in violent motion). You will note it performs all sorts of antics without my getting near it. This simply shows the influence of one piece of iron upon another in an alternating magnetic field. I will now take this ring of copper and bring it over the coil. You will note that it is repelled and held up without visible support. I will hang it on this support at one side and you see it is repelled and supported in the air. (Fig. 11). The magnetic waves are trying to get through the ring but the ring is producing counter waves and backing them up or reflecting them. The waves have to go out sidewise and in changing their course they hold the ring up, repelling it with such force that the ring floats as it were in the air. (The lecturer took another ring of the same size and

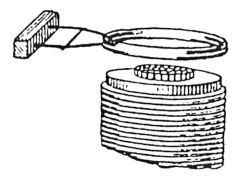

FIG. 11.

placed it underneath the first. The upper ring supported the lower in the air, the two being powerfully attracted. These two rings tend to attract. There the second ring is sticking to the first. Both of these rings are affected alike. They sympathize with each other in a sense. They have currents developed in the same direction in them by the magnetic waves striking through, and having currents in the same direction they are attracted. Were the currents in opposite directions they would repel each other. (The lecturer then held up a bottle partly filled with water with an incandescent lamp inside attached to the terminals of a coil of wire, thoroughly insulated, the lamp coil being entirely submerged, Fig. 12).

In this bottle we have a little coil of wire connected to an incandescent lamp and placing it over the coil and core thus the incandescent lamp is supplied with current, and the lamp and coil are floated up a little in the fluid by the repulsive action tending to push away the little coil. This proves at once that the repulsive actions shown are due to currents, for you see we have the lamp lighted by currents produced in the coil. The lamp is lighted underneath the water. As it rises the light given by the lamp gets dimmer. The waves are at greater distances and more spread out. In that way the lamp is less brilliant. I am indebted to Mr. Heinze for getting the lamp into the bottle through that neck. (Laughter). (The coil and lamp were at least three times the diameter of the neck of the bottle).

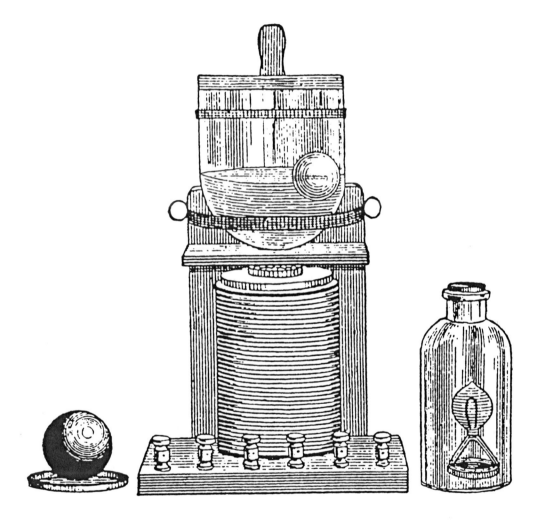

FIG. 12.

Here is an ordinary incandescent lamp and I will bring the coil to which it is attached into the field (Fig. 13). I get the same effect as I did with the bottle. We are producing these waves in the ether around the coil at the rate of 250 impulses per second, sending them out from the coil, meeting them by another copper coil, and converting them back again into current, as you see. The waves do not require air because we can, as in our bottle experiment, use water. They went through the glass, through the bottle, and through the water. We might have used other things and they would still have gone through. It is not the water that is concerned in this operation. They go through solid glass. They go through solid pitch. Can these waves go unrestrictedly through everything? No. The waves of light go through glass. These waves went through glass also, the electric waves. They went through water, so does light. But light does not get through that solid piece of copper. I will see whether these waves go through it. These electric waves do not go through the copper plate.

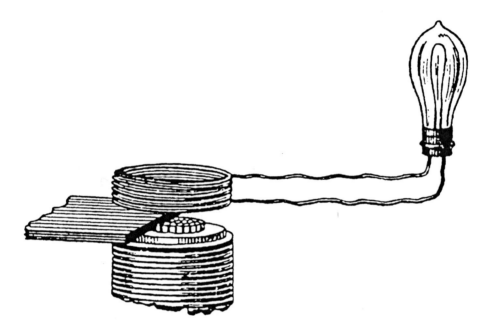

FIG. 13.

(The incandescent lamp and coil were used, the coil over the magnetized core with a sheet of heavy copper intervening). We can turn the lamp off and on by this novel process of shielding the waves. By removing the copper or replacing it in part, we can make it bright or we can make it dull. Do not say then that we can not turn our electric lamps up and down as we can gas.

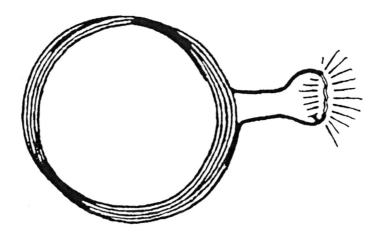

FIG. 14.

Another demonstration is to take a piece of platinum wire and attach it to the terminals of a small coil (Fig. 14). I will then take that coil and bring it within the influence of the magnetic waves and we will see what we get. (Experiment tried). The platinum wire is heated. It is giving out more heat than light. But I can do too much of that (increasing the intensity of the light or incandescence by approaching the coil quite near the core). I can put an extension on the iron core of the coil below and then bring down the coil and I at once melt the platinum wire.

I have here another little coil and its ends are sticking out. (A coil of coarse wire was shown with its ends projecting outwardly and parallel to each other.) (Approaching this coil to the core). Nothing occurs because I have not connected the ends. If I do connect the ends solidly the coil will jump off. (Doing this). There, it has jumped off. I will now close the circuit of the coil through these pieces of iron wire. I have welded the two pieces of wire together. I can repeat that operation with different wires. I can even bend the wire in a circle, bring the ends together, and make it into a closed loop. (Experiment accomplished). We have constructed here then a very simple electric welding machine.

We can further show that these waves emitted are capable of producing other actions. (Producing a copper ring within which was placed a copper ball, and the two placed over the magnetized core on a mica plate). Here the waves are actually pushing the ball around. (The ball was spinning rapidly). I have also here a brass plate which I will use instead of the copper ring under the ball and we will see what is the effect on the brass ball. It rotates as before. (Changing the position of the ball). I think it will stop and reverse. You see it does that promptly. Suppose we keep it in the middle. (The ball rolls away). It does not seek to be there. (Taking a copper dish and placing it on the plate). The dish spins. It is bound to do the same thing as the ball. (Taking up a small ball). We will try whether the children of this electrical world will do these things. Little people it is said go a great deal faster than the big ones. The small ball spins and dances about rapidly. Like all children it is more erratic in its ways. Here is the old grandfather. (Using a large ball which moved slowly). He seems rather slow about getting into that frivolous condition of waltzing as it were, but they all come to it. See how dignified he is about it. (Laughter). We will show that this action is not affected by the air at all because we are going to remove the air. We are going to put in place of it a quantity of water and see what we get. (A glass vase partly filled with water was suspended over the magnetized core, Fig. 12). Now we will put the ball in this glass vase with the water. There, the ball rolls over and over in the water, spinning rapidly. We can greatly enhance this effect by putting a ring of copper in a certain position. (Placing a copper ring about the ball at an angle). Now, the ball picks up and goes at a much greater rate of speed. We may take the large ball instead (substituting) and the effects are very much the same. (The large ball began moving quite rapidly). Grandfather prefers water. (Laughter). (Placing the copper dish in the water). The dish is rotating rapidly and to show that these waves are not by any means exhausted by these operations we will put the ball in the dish. There, the dish, ball, and little ball are all turning around in the water. You will notice that in these experiments I put this copper plate underneath the water. The waves must be modified. We must put this plate, which is not perfectly transparent to the waves, in one part of the magnetic field and shut off some of the waves, so as to let them go out freely in another part, and then those rotary actions occur.

I have here a disc of copper mounted so that it can rotate. Here is another which is of iron. Copper is a non-magnetic substance, but it is an excellent conductor of electricity. Iron is not so good a conductor but it is magnetic. I will first put this copper shield below the disc. You will notice that the disc is now rapidly rotating, but if I should put the copper disc there without the shield under it there is found no such action. Under these conditions neither does this disc (the iron one) turn, but suppose I take these two discs and bring them into the same field, one is magnetic and the other not. (Copper below the iron.) Now both discs are in rapid rotation and they are going in opposite directions relatively. The iron disc is turning to the left as I look at it and the copper disc to the right. It is just as if they were geared or belted together. Here is a steel file (placing the file flat on the magnetized core). I will take this iron disc and put it over the file. Now, it is beginning to turn. The file is the cause of the turning now, it has so modified the waves as to cause them to turn the disc. The disc is set in motion by the file. The copper disc likewise (substituting) is set in motion in the same manner as the iron. Now I put the file in position and apply both discs and the iron disc is going in one direction and the copper in the other, the directions of motion have been reversed from what they were without the file. They are going like geared wheels now, but not of course together. I can apply the iron disc to the file even in this way in a vertical plane and the rotation begins at once. (Substituting.) The copper disc applied in that way gives no effect. Copper differs from iron in not being magnetic and does not turn in this position, but lay it down horizontally near the file and immediately the copper takes on the rotation. (Placing the copper flat and the iron at right angles thereto, both start rotating, Fig. 15). Here we have an example of crown gearing. A different experiment is to take another core of iron wire and hold it above the end of the magnetized core and place the copper disc between in a selected relation. Immediately the rotation begins and gets up to a very rapid rate.

FIG. 15.

(Going to another piece of apparatus, Fig. 16.) Here I have a copper disc arranged on a stand and moving in a slot in an iron ring in such a way that we can actually develop considerable power by connecting this coil on the ring into the alternating current circuit. The disc is now in rapid rotation and exerts power. Hence this is an alternating current motor made on a similar plan to the preceding experiments. (Pointing to the slot or space between the magnetic poles after the disc was removed.) There is where the disc was. I will take this copper disc (the one mounted on a handle) and put its edge in the slot. It is in rapid rotation as was the other. I can balance it on my finger. (The lecturer supported the outer end of the handle by placing his finger under it and the disc remained in suspension between the magnetic poles while turning rapidly.) It not only propels the disc but supports it at the same time. This apparatus would make an excellent counterfeit coin detector, a "nickel-in-the-slot" apparatus. Lay a coin here on a little shelf, such as a 10 cent piece, and we will see "how the thing works." It disappears the moment you let it go. The tendency is to pull the coin through the slot. It will take from me a silver dollar and throw it vertically (illustrating) clear above the machine, (Fig. 17). I would say here that you could not get it to do that with a lead coin. (Applause.) The machine knows the difference. Silver is an excellent conductor of electricity. Lead is so poor a one that it would hardly be affected in that slot at all. The machine is very choice in its selection. It takes silver every time. It might take gold, but it is constituted to

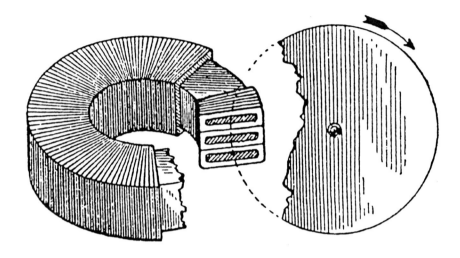

FIG. 16.

show a preference for silver. The machine takes silver and copper because they are both very good conductors of electricity, and are therefore strongly affected by the electric waves, but what is silver in relation to light waves, what is silver in relation to waves of extreme rapidity of vibration? It is a fine reflector. It is the best reflector of light waves we have. Silver will turn back the rays of light very perfectly. So a silver plate as well as a copper one will answer in shading off electric waves, in preventing their passage. It would reflect them, drive them back, not let them get through. It behaves very much the same with light waves. Light strikes a silver mirror and back it goes on a reflected course.

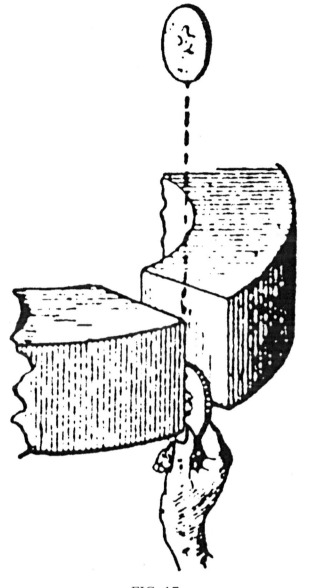

FIG. 17.

Here I have a cast iron ring with a small copper coil wound on it at one place and the ends of the coil disconnected, so that no current can flow in it. I put the pivoted iron disc over the ring parallel to it so as to get it about the center or concentric. The iron ring is laid on the alternating magnetic coil at a part at one side of the coil on the ring. (Fig. 18). I close the circuit of the small coil and we see what occurs. Now the disc begins to turn. That simply shows that the copper coil reflects--sends back the magnetic waves that want to go around the

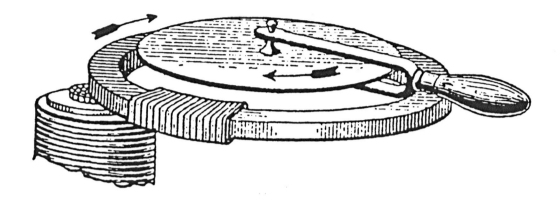

FIG. 18.

ring in this way (indicating), while on that side of the ring they can go around. They carry the iron disc around with them. Each of these experiments show the existence of something in the ether. We can substitute different substances in the space but we can not put metal objects there without modifying the results.

Now, what bearing have these experiments on practical work? We run our electric lights by what are called electric currents. We run our electric motors by what are called electric currents. We run our street cars by electric motors driven by electric currents. What is that electric current? It is an effect in the wire, accompanied by a transference of what we call energy. The electric current is only part of it, and that is the main point in which modern ideas differ from those of past times. The wire is not the principal thing conveying the energy. If we had not around that wire a medium which could be filled up with waves of magnetism, or with a strain in it, no power would be conveyed by that wire, or very little. The actual fact is that the wire which conveys current is merely a core of the medium in which the magnetism exists, the thing which really carries the power. In driving a car it is not the current in the wire which is doing the work, although that is the center of the action. All around the wire in the ether-- not in the air because that can be taken away-is a set of disturbances called magnetic, which proceed along the wire outside of it until they get down into the motors underneath the car. The motors are merely well organized devices for receiving these magnetic strains and converting them into power, for converting them just as the steam engine which drives the dynamo converts the heat motion into power. We have then in the electric distribution of power the change complete from heat to power, from power to electricity and back to power again.

I have also shown you that light is due to a similar action, that it is caused by electro-magnetic waves or changing strains in the ether whose rate of motion is about 500 trillions and more per second, that we can say when we see anything, we see by electro-magnetic actions. The sun then, shining on the earth is nothing more nor less than a great electric wave generator, as it were, but it is not generating such waves as we have been experimenting with here and sending them out, but waves having a rate of 500 trillions a second and more, travelling between the sun and the earth at the rate of approximately 200,000 miles a second. These waves we have used traverse the ether also at the rate of approximately 200,000 miles a second. The velocity of that wave (referring to the magnetic waves produced by the coil) has been shown to be just the velocity of light, which is another thing to prove that the two are the same in nature. The movements are made in the same manner. Their actions of reflection and refraction are alike. Silver which is a splendid reflector for these waves is a splendid medium for throwing back the light waves. Glass, which has no influence in stopping these waves is also transparent to light and is a splendid electric insulator. It does not allow electricity to flow in it. It does not beat back the waves as silver or copper does, but it is just like the ether, transparent to the magnetic, electric and light waves. Whenever we have a black substance like carbon we see that it does not let the light get through. It does not reflect back the waves, but absorbs them-most of them. If the waves from our coil were greatly increased in rapidity they would not get through a plate of carbon. Yet very few would be reflected. It is not a good enough conductor to act as a reflector but would let the waves into its black surface-absorb them and convert them into heat. If held up to the sunlight the sun's rays enter and are converted into heat. I want to bring absorbed heat into this category too--it has not been done but it can be done legitimately. I say that if the waves strike that plate and it is of polished silver, they are reflected--they do not enter, but I am not quite right. If the waves are like those we have used, of slow rate, they will enter in part and they will only be partly reflected. The silver or copper plate will get hot in a short time when strongly influenced by these slow magnetic waves. It will have absorbed a certain amount of the energy of waves and converted it into heat. But if I put that copper or silver plate in the sunlight for a time it is also heated to some extent. The electro-magnetic waves striking the plate and which are not reflected set up currents of electricity in the plate, and these currents are resisted in their flow by the copper not being a perfect conductor. If it were a perfect conductor it would not heat and would be a perfect reflector. The particles are, as it were, all set in motion by the current flow--they shake. They get more and more into a condition of shaking as the copper gets heated. The waves of sun light are small in extent and we can not expect them to produce tangible electric currents in the copper on absorption, but I believe the action is none the less that the plate heats when put into the sun's light because the electro-magnetic waves traversing the copper have produced intensely local electric currents in or between the molecules of the copper, which currents heat the plate.

I could go on still further amplifying upon this interesting subject, but I have taken up quite enough of your time this evening and trust that I may have interested you. It only remains for me to thank you for your attention and patience while I have been addressing you. (Prolonged applause.)

GEORGE WESTINGHOUSE was born in Central Bridge, New York in 1846 and died in New York City in 1914. He was an inventor and businessman. As a businessman, he was the major factor in the adoption of alternating current electric power transmission in the U.S. He was interested in the railroad for which he created his best known invention, the air brake. Westinghouse created more than 100 patents.

Following is a speech given by George Westinghouse before the Southern Commercial Congress at Atlanta, Georgia, March, 1911. It appears as a courtesy of the Westinghouse Museum, Pittsburgh, PA.

"ELECTRICITY IN THE DEVELOPMENT OF THE SOUTH"

If we examine broadly the changes which have come about in industrial methods and in the means of transportation since the invention of the steam engine, it will be found that the application of power has been the fundamental factor in bringing about the characteristic conditions of the era in which we live. The steam vessel and the steam locomotive, by revolutionizing transportation methods, made possible the present development of our country. It is the power of the steam engine or the waterwheel which has substituted the power loom for the hand loom, with all the marvelous results which have followed. Similarly, throughout nearly every industry, human muscle is no longer the source of power, for the hand now directs and controls the untiring and unlimited power of great engines. Reduced to its ultimate terms, the vital forces in industry and in transportation come from coal mines and waterfalls, resources with which the South is abundantly blessed, and the problem is to secure power from these sources and to utilize it in building up the industrial and commercial life of the community.

The Age of Electricity

Had Jules Verne sought to imagine some universal servant of mankind, he would well have depicted some magic agent which would apply Nature's forces to do man's work; which could take the energy of her hidden coal, of the air, or of her falling water, carry it by easy channels and cause it to give the light of a million candles, the power of a thousand men, or to move great loads faster than horses could travel, to produce heat without combustion, and to unlock chemical bonds and release new materials. No wonder such was pictured by the imagination of the seers of the past; and yet a subtle force which transcends the powers of the imagination is daily doing all these things--a vitalizing force, which is already stimulating the physical recovery of the South; and if we still think of the present as the era of steam and steel, unquestionably the coming epoch, whose dawn we are privileged to witness will be known as the Age of Electricity. First the toy, and long the mystery of the scientist, electric power is now a familiar tool for the accomplishment of the work and the increase of the comfort and pleasure of mankind.

Although we may not know the ultimate nature of electricity, yet we do know some of its essential laws and methods of controlling and using it.

During the twenty-five years in which I have been intimately interested in the electrical art, a development has been witnessed which surpassed the most optimistic predictions. At the beginning of this period it was the general conviction that electricity would be limited to local use in the lighting of densely populated districts or the supply of power to adjacent factories. Indeed, there had been no developments to remotely foreshadow what has since been accomplished.

<u>A Simple but Great Invention</u>--At that period, however, there had already been developed and operated electric arc lighting circuits of high voltage, extended over rather large areas, with the pressure upon the wires of from 2000 to 7000 volts, which practically demonstrated that considerable electric power could be cheaply transmitted if means could be found to utilize safely high-voltage electric current for power and light and for other purposes; but such means were not then known. It often happens, when something is greatly needed for any great purpose, that as a result of a lively appreciation by many of the existing need, there arises in due course invention or discovery which meets the demand, and so it was in the matter of invention and discovery which gave us a simple static device, consisting of two coils of copper wire surrounded by sheets of iron, which could, without an appreciable loss of energy, transform alternating electric currents of high-voltage and small quantity, dangerous to life, to low-voltage currents of large quantity, safely available for all power, light, heat and other purposes.

To the part I took in bringing forward in the '80's of the last century the <u>alternating-current system</u> of electric generation and distribution, I owe much, if not all, of the reputation accorded to me as one of the many pioneers in what is now a great and important industry.

<u>When Restrictive Laws Would Have Defeated Progress</u>--The introduction of alternating-current apparatus was bitterly opposed by those who were then exploiting direct-current apparatus, and legislation was sought to prohibit its use because of its alleged danger to life. I mention this incident because it clearly shows that <u>restrictive</u> laws are not always advantageous, for had the legislation sought for by the opponents of the alternating current system been secured and enforced, I would not now have any justification for this address, because the influence of electricity in the development of the South would be too unimportant to entitle it consideration on this occasion.

Electric Power Carried Over 200 Miles--As a result of the development of the alternating current and of years of experience in the manufacture of electric transformers and of insulators for supporting electric conductors, power is now successfully transmitted by alternating current over distances of 200 miles or more. Thus water power in almost inaccessible places awaits only the coming of engineers and of capital to be made available for industrial purposes.

It is estimated by those who have made a study of the sources of water power of the Appalachian Mountains, that there can ultimately be developed from 5,000,000 to 7,000,000 horsepower during the dry season of the year, and a much larger quantity at other times. This great water power is brought by Nature to your mountains and hills in widely varying quantities and will continue indefinitely; but the maximum and minimum flow of the waters of your rivers can be affected by the works of man and by a wise conservation of your forests.

Electricity to do the World's Work--Notwithstanding our familiarity with the present uses of electricity, few of us really comprehend how universal and fundamental is the part which electricity is destined to assume in the life of the future generations. Nothing else can convey, distribute and apply power in a way which compares with electricity. From one dynamo can be taken the power for operating the telephone and the telegraph, the power for lightning, the power for operating street cars and railroad trains, the power for operating mills and factories and mines, the power for electro-chemistry, the power for heating. Electricity is a universal means of applying power for doing the physical work of the world. It is effective, not only in the application, but in the production of power. Less coal is required for producing electric power on a large scale than is required when many individual engines of smaller size are used. Water powers which otherwise would be unavailable are made useful for supplying power to distant cities, and even a mill located at a water power will give better service when it uses the electric drive. Electricity affords a simpler, better way of doing many things with which we are familiar, and it also makes possible new methods and new developments which, without it, would be impossible.

With electric power the mill can draw its energy from any stream within a radius of a hundred miles or more; it may be located on high and healthful ground, on the outskirts of an established town or city where labor is plentiful and transportation facilities are the best. In the plan and design of the mill itself, there is no longer the necessity of arranging buildings and machinery to be operated from great belts and long shafting taking power from a single source; but individual motors in each department, or on each machine or loom enable the whole plant to be laid out so as to give economy in construction, convenience in handling materials, and ensure the safety and health of employees, thus securing a freedom and an excellence which is impossible without electricity.

The oppressive heat of the summer months in the South can be made tolerable by cooling devices and fans operated by electricity, and electric heaters, which are always ready for instantaneous service, can be used during the short intervals in the winter when artificial heat is necessary for comfort and health.

Conservation of Coal Resources--Furthermore, the use of electricity will conserve the coal deposits of the world for those industrial processes in the performance of which it may always be an indispensable element. To illustrate what a conservator of the coal resources of the country water power may prove, I will only mention that to produce for ten hours each day from coal the five million horsepower which may be developed from Southern water powers, would require, with the most efficient kinds of engines, not less than 25 million tons of coal annually. If there were no water power available, methods would be adopted for producing power and conserving heat, which would effect a saving of over one half of the coal now consumed in the world. Here is a field for agitation against waste of our natural resources surpassing all others in importance.

The South's Opportunity

Now, what is the significance to the South of these facts? How can the South, which has almost everything before it in the matter of industrial affairs requiring the aid of modern achievement, by the foresight and by promptly grasping the opportunities which are presented to it, hasten its industrial development, increase its wealth, improve the health of its people and increase their happiness?

Truly, here are subjects not to be circumscribed by the wisdom and judgement of one man, but calling for the united counsel and effort of the wisest and best among us--requiring not merely the knowledge of the scientist, the skill of the engineer and the wealth of the capitalist, but also the broad view, the enlightened experience and the high endeavor of our greatest statesmen.

Present Achievements in the South--In the development and utilization of the energy of waterfalls, the South has already taken a leading position, and the industrial benefits thereof are so widely and favorably known that no argument is now needed to justify the work already done or to point out the great and lasting benefits to be derived from its extension.

Any address on electricity in the South would be incomplete without an expression of high appreciation of the work of the Southern Power Company, begun by Dr. Wylie and developed to its present stage by the Messrs. Duke.

A Great Electrical System--This is the largest power transmission system in the South and is among the most extensive and important in the country. It is not a simple transmission line from a single power house to a single mill or city, but an extensive system which receives power from many power plants on different streams in several States. Hence, low-water or high-water on one river, which might temporarily disable certain plants, has but a slight effect on the whole system.

The lines of the Southern Power Company extend 150 miles north and south and 200 miles east and west, and connect into a single hydroelectric power system plants aggregating 100,000 horsepower. It is a magnificent demonstration of what electricity can do to conserve and utilize water power in developing the great and growing textile and other industries of the South. The Southern Power Company is furnishing light to 45 cities and towns and supplying current to 6 street railway systems and to hundreds of motors for various uses. This power development is the result of intelligent and far-sighted business courage and confidence in Southern affairs, which have inspired and actuated the men who have built up this great enterprise.

I am informed that the millions already invested in the Southern Power Company have not yet yielded even a moderate net income to those who have put their money into an investment which has benefited others more than themselves by insuring an increase in production and profit to its patrons, a striking evidence of the importance of a generous treatment by authorities as well as by those who derive an absolute money benefit.

Industries Likely to be Developed--The industries most likely to be developed and to increase because of peculiar suitability to conditions now existing in the South are:--textile mills, fertilizer works, cement plants, coal, iron, copper and gold mining, ore reduction plants, iron and steel mills, agricultural implement works, canning factories, road building, furniture manufacture, lumber plants, paper mills, shoe and leather factories, and oil refineries, in all of which industries electric power increases production.

Electricity in Metallurgy--The South abounds in coal and iron as well as other metals, which can be cheaply mined. Owing to the presence of impurities in the iron ore, especially phosphorus, the pig irons produced in the South have not been considered so suitable for steel manufacture as those made from the purer ores of the North. The electric furnaces for refining steel, which have been recently developed and quite extensively used, will make available the iron resources of the South in the production of the high grades of steel, and it is no stretch of imagination to foresee that the South will become a large producer of the raw material, and through the cheapness of its labor will be able to turn these materials into finished products. At the same time the slag by-product of blast furnaces will remain to be used for fertilizing purposes.

Electrical Production of Fertilizers--The South is already a large user of fertilizers, much of which is imported and the supply of which is limited and exhaustible, nitrogen forming an important part of the fertilizers which are commonly used. During the past few years great attention has been given to the development of means for the electric production of fertilizers and so much has already been accomplished that it may be said with confidence that the fertilization of our soil within the near future will be largely dependent upon electricity. Most of the material required, coal and limestone, for this purpose, is found in the South in unlimited quantities. Were the soils in the United States as carefully tilled and fertilized as in many densely populated countries, there would be an immense increase in our agricultural products.

Electricity in Cotton Mills--A brief consideration of the special advantages already derived from the use of electric power in the cotton industry will well illustrate the benefits to be gained from the general extension in the use of this wonderful force to other fields.

The output of cotton mills has been increased and the quality of goods is improved, due largely to the uniform speed attained by the electric drive compared with power conveyed through belts and lines of shafting. This uniform speed has resulted in an increased production with an increased profit, which in some cases exceeds the cost of the electric power. With electric drives, recording meters can be placed in the circuits which supply power, and the instantaneous power or the total power for any given time can thus be ascertained, a feature of great value to the management in determining whether separate departments of the mill are starting or stopping on time and whether the full load is kept on the machines during working hours.

With electric drives, one set of machines or a part of a mill can be independently operated when it is not advantageous or convenient to run the whole mill. When there is a single power house with mechanical drive, any enlargement must be conditioned upon the extension of shafting or belting; but with electricity, wires can be readily run to any point in the old buildings or to new buildings.

In the territory of the Southern Power Company, it was at first difficult to induce the mill managers to adopt electric power, and it took three years of effort to introduce ten thousand horsepower; then, however, the mill managers observed the advantages of their neighbors who used electric power with the result that at the end of the next period of three years electric power had increased to more than 65,000 horsepower, while now there is a total of 80,000 horsepower of electrical machinery installed.

Of the 300 or more cotton mills in North Carolina, about 25 percent are now wholly driven electrically. Although there has been a great increase in the number of cotton mills in the South in recent years, the mills have been devoted to the production of the cheaper grades of cloth; but it is predicted that the future growth will not be merely in the number of mills, but will be in the production of the finer grades of cotton fabrics.

In Transportation--I have briefly sketched the fundamental place which electric power distribution is taking in industrial activities, and I have briefly referred to what one electric power transmission company is accomplishing in pushing the textile industry in which the South takes just pride. Time does not permit me to catalogue all the possibilities of electricity in the development of this great country. The South has mineral resources to be developed--electricity is the established method of mining operations. The main railway lines of the South run north and south--electricity enables trolley lines to be run east and west to serve as feeders for the trunk lines, and when electricity is used for the operation of your railways, as it will certainly be some day, there will follow a more intimate relationship between producers and carriers than might otherwise exist.

Looking Forward

Having been asked to speak upon the subject of electricity in the development of the South because of my connection with the electrical industries of the country, it seems to me I cannot fulfill the expectations of those who have planned this Congress by limiting my observations to matters with which you are more or less familiar from personal experience or from articles in your daily papers and in magazines; I should also ask you to look forward to what we may expect in the years to come.

Electricity in Agriculture and Horticulture--In 1906-1907 some experiments were made in England with the cooperation of Sir Oliver Lodge, the eminent English scientist, in the stimulation of plant growth by electricity. It has been frequently observed that plant growth is stimulated by electric light, and numerous experiments have been made having for their object the stimulation of the soil by the application of electric current. The experiments reported by Sir Oliver Lodge in a privately printed brochure on Electricity in Agriculture are briefly as follows:--

Two tracts of land about twenty acres each were similarly sown or planted. On half of this land poles with insulators were erected to support the electric wires, only one pole per acre being required for the purpose. The electricity required was produced by a small dynamo driven by a two horsepower oil engine and was transformed to a tension of about 100,000 volts of very high frequency. The experiments, which extended over several years, give remarkable results, an increase of from 30 to 40 percent being secured in wheat crops grown on the electrified plot as compared with the crop produced on the unelectrified plot. Moreover, the electrified wheat was of a better milling and baking quality and sold at considerably higher price than that grown on the unelectrified plot. Similar experiments with strawberries, marigolds, tomatoes, cucumbers, beets and carrots showed equally remarkable results. One-year strawberry plants showed in one instance 80 percent increase and more runners produced, while with five-year plants the increase was 36 percent.

In writing to me on this subject in response to my request, in order that I might make a reference to it in this address, Sir Oliver Lodge suggested that the results attained in the experiments referred to and in others would justify an elaborate series of experiments. These experiments could be usefully undertaken at the stations under the control of the Agriculture Department.

An explanation given for the excitation of vegetation by these high tension currents is that high frequency electrical discharges favorably affect the deposit of the nitrogen in the atmosphere into the soil, upon which deposit vegetation so largely subsists.

Whatever prevents disease and ensures health contributes not only to man's happiness, but also to his efficiency, and it appears that the electric current is to play a very important part in this field.

Mercury Vapor Lamps--Ultra-Violet Rays--The outcome of the efforts of one who specializes in any particular kind of apparatus is often interesting. The development, by Doctor Peter Cooper Hewitt, of the mercury vapor lamp, has provided a light which is the least fatiguing to the human eye of all artificial lights, and experimentation with this lamp has led to the development of several other uses of the mercury vapor arc, one of which is the production of quartz tubes of ultra-violet rays, the effects of which are likely to be of the very highest importance in our daily lives. While these ultra-violet rays are emitted in the quartz tubes, they are effectively neutralized by the glass tubes which contain the mercury vapor used in lighting.

Sterilizing Water and Milk--One of the important uses to which these ultra-violet rays have already been put has been to absolutely sterilize water, however much it may have been contaminated by bacteria. Experiments have also shown that the ultra-violet rays will sterilize milk without the application of heat in such a manner that it can be kept in properly sterilized vessels for long periods without deterioration or loss of its food values.

With the growth of population, the pollution of rivers, and the contamination of the water supply upon which our population must rely, and the difficulty of determining whether the water and milk we use are free from noxious bacteria, this safe and thorough method of sterilization becomes of estimable value. The elaborate experiments and demonstrations which have already been made at the Sorbonne, in Paris, and at the City Water Work of Marseilles, France, have not only proved the feasibility of this method of sterilization, but have brought out the fact that a 15,000-kw generator of electrical energy could sterilize, by means of mercury vapor quartz lamps, as much water as is actually used for drinking and cooking in the United States.

The simplicity of the apparatus for sterilizing water is such that there is no doubt but that it can be advantageously installed in factories and other places, and even in dwellings, adjacent to the point or points where the water is to be used, thus avoiding any possible contamination between the point of supply and point of use.

The electric energy required for the operation of a quartz mercury vapor lamp used for the daily sterilization of 85,000 gallons of water is about equal to that required for half a dozen ordinary incandescent lamps.

Ageing of Wine--Not only have water and milk been sterilized, but in other experiments, also carried on at the Sorbonne, it was found that new wine was affected in a manner to give it the qualities normally attained in years, or an age of apparently many years was given by a few second's application of the ultra-violet rays.

These experiments and investigations suggest that uses for the ultra-violet rays will be found which have not yet been conceived.

Supplanting Costly Apparatus--An important use of the mercury vapor apparatus has been to transform or rectify alternating currents into continuous currents, and some recent experiments indicate that this can be done on a large scale with a considerable saving of electrical energy. These promising results foreshadow the disappearance of the costly rotating apparatus which is now used for that purpose in the operation of railways and for purposes where the use of a continuous current is advantageous.

Hertzian Waves--The transmission of electrical energy through the atmosphere without wires has, in a very few years, so far advanced that wireless telegraphy is now an important feature of our daily life. We read of instances where wireless messages have been received at a distance of over three thousand miles from the point at which they were sent, and it is said that we shall shortly have regular wireless communication between Paris and New York.

Portable Wireless Telephones--Not only has it been possible to communicate by wireless in the Morse code, but it has been found that, with suitable apparatus, telephone conversations can be carried on over considerable distances, and it is expected that by improvement in the apparatus, conversations can be carried on over very considerable distances. Investigations, of which there is almost daily mention in the public press, indicate such great simplification in wireless telephone apparatus that we may, within the quite near future, have placed at our disposal a simple portable apparatus which will permit wireless conversation to be carried on over a considerable area. This will prove of great value in sparsely settled districts.

Possibilities of Hertzian Waves--It may interest you to know that the frequency of the electrical waves sent out by some forms of wireless transmitters approaches a million per second, and that by either an increase in the amplitude of these vibrations or by a more sensitive receiver, the distance over which these waves (which undoubtedly extend to an infinite distance) may be recorded, can be greatly increased.

In an experiment made by Doctor Peter Cooper Hewitt with powerful wireless transmission apparatus, including a mercury vapor interrupter, it was found that the effect of the high-frequency discharge upon the iron in the building occupied, such as water and heater pipes, quickly produced incipient fires within the room where the apparatus was erected, thus demonstrating the wonderful power of this incomprehensible force and suggesting great possibilities in the transmission of electrical energy without wires.

The transmission of electric energy without wires which will be especially valuable for signalling purposes and for the control of machinery at a distance, will undoubtedly play a most important part in army and navy operations.

<u>Lord Kelvin on Radium</u>--We are hearing and learning more and more in regard to the power of radium, and predictions have been made that it will some day furnish power in great quantities. This I very much doubt. The popular belief is that radium constantly produces heat and light without and appreciable loss in its weight, and that it will continuously produce heat. Lord Kelvin, whom I had the honor of knowing, was greatly interested in the discovery of radium by Madame Curie. In one of the last conversations I had with him, I ventured to give a conception of the cause of the "production" of heat by radium, my idea being that the radium acts as a transformer of one of the forces of ether into some other form of force, and that in such transformation heat is produced. Lord Kelvin, who had studied the subject, said that he had already arrived at the same conclusion on the general hypothesis that neither heat nor light can be produced without energy. I refer to this because of the indication that there exists a form of energy of which we have as yet no knowledge, but which may yet become available to us as a result of further discoveries.

Cooperation Should be Compulsory

The advantages of cooperation in the matter of the development and supply of electricity, having regard to a lessening of the cost and insuring the certainty of supply, cannot be overestimated, and those already secured by operations on a large scale are well known. Further cooperation in this great work for the benefit of the public, if not voluntary in the future, should, in my opinion, be an enforced one, notwithstanding the outcry which has been raised by the ill-informed and with reference to an imaginary monopolization of the water power of the nation. Encouragement should be given to the investment of capital in the development of these enterprises under such wise and reasonable regulation as will insure economy in the construction and operation of plants, adequate returns to the capital invested, and at the same time protect the consumer against exorbitant rates and charges or unfair discrimination.

In the larger industrial developments which I foresee for the South there are other important factors which equal in importance the development of the water-power resources upon which I have dwelt. I have particularly in mind those existing restrictions which make it difficult and expensive for a small corporation to carry on conveniently and in a simple manner its business with ramifications in several states, which restrictions, however, the great corporations of the country can easily surmount by reason of their financial ability to organize separate subsidiary companies in those States where such an expedient is rendered necessary to meet legislative requirements.

Federal Incorporation

I have long held that a Federal Incorporation Act, which the President advocates, under which all companies doing an interstate business could organize, would be a solution of the difficulties which are now almost insurmountable, and which are being added to in an alarming manner in the endeavor of the legislatures of the several States to curb a few of the tens of thousands of companies and firms doing an interstate business.

Protection of Minority Owners--After having read and carefully studied the bill providing for Federal incorporation, which was introduced in the long session of the present Congress, I am constrained to say I would prefer to see a Federal law in terms more easily comprehended by business men and devoid of those provisions which would give a privileged few a practical control of a corporation by expedients which have been skillfully developed and which are now looked upon as a matter of course.

I have in mind particularity the depriving of minority owners of possible representation by the formation of voting trusts and the election of directors in classes, methods which can, and often do, defeat the purposes of laws which have provided for cumulative voting, whereby a substantial minority can insure the election of at least one member of a board of directors.

Directors Should be Large Shareholders and Elected Annually--In my judgement, each director of a corporation should be required actually to own a substantial interest in the shares of the company, the affairs of which he aids to control, and the term of office be only from year to year. To make my meaning clearer, I will illustrate by supposing that a company had, by appropriate bylaws, established a board of five directors, only one of whom could be elected each year. Obviously, the provision of the law for cumulative voting would have no meaning in the government of the affairs of such a company.

It may be unorthodox to say this, but it is my conviction that the conduct of a business without profit is disadvantageous to the community at large because of its demoralizing effect on the industry and its influence upon others. A Federal Incorporation Act should provide for a statement, on prescribed forms, of the assets and liabilities of each corporation taking advantage of its provisions. This statement should be available to all who are asked to extend credit to the corporation. The disadvantage to a company of doing business at a loss under such conditions need not be enlarged upon.

Each of the great corporations of to-day had its origin in a business established by an individual or small company based upon the skill and efforts of one or more individuals. The development of the South must be more or less rapid according as the work of such men is appreciated and encouraged, especially during the period of strenuous effort necessary to the building up of large and prosperous industries from small beginnings.

The Training of Young Men

In conclusion, I urge the young men of the South to make themselves familiar with industrial affairs by learning to be proficient in the use of their hands as well as in the use of their heads. My early greatest capital was the experience and skill acquired from the opportunity given me when I was young to work with all kinds of machinery, coupled later with lessons in that discipline to which a soldier is required to submit, and the acquirement of a spirit of readiness to carry out the instructions of superiors. President Taft's statement that the introduction of military discipline in the schools and colleges of the land, in the advantages of which all would participate, would be of greater benefit to our country than the high development of athletics by a few, is worthy of most serious attention. The present preeminence of Germany in industrial matters arises very largely from the military training and discipline to which each of her citizens must submit.